T0228587

BURBOT

BURBOT

Conserving the Enigmatic Freshwater Codfish

Dr Mark Everard

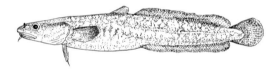

5m Books

First published 2021

Published by
5M Books Ltd,
Lings, Great Easton,
Essex CM6 2HH, UK,
Tel: +44 (0) 330 1333 580
www.5mbooks.com

A Catalogue record for this book is available from the British Library

ISBN 9781789181456
eISBN 9781789181616
DOI 10.52517.9781789181616

Book layout by Servis Filmsetting Ltd, Stockport, Cheshire
Printed by Bell & Bain Ltd, Glasgow

Front cover: Batbold Tsedensodnoml Johan Auwerx
Back cover: Joachim Claeyé; Hendrik Wocher; Ian Wellby

Contents

Acknowledgements

The drafting of this book has benefitted from the input of many colleagues and experts and the encouragement of friends for what is, by any standards, a bit of a mad venture!

In particular, I would like to thank Jonah Tosney (Norfolk Rivers Trust), Ian Wellby (Blueroof Ltd) and Tom Worthington (Cambridge University) for their much-valued inputs and support. An 'absent friend' in this regard is the late Keith Easton, former Fisheries Scientist with the Environment Agency, who had been a lifelong cheerleader for the burbot. Keith was an instigator of Tom Worthington's PhD on potential burbot reintroduction and of Ian Wellby's involvement too. In Ian's view, we would not be where we are today with burbot reintroduction into the UK without Keith, his early work and his encouragement of others.

My thanks also go to Johan Auwerx of INBO (Instituut voor Natuur-en Bosonderzoek: the Research Institute for Nature and Forest) in Belgium, who has shared with me his deep experience of the burbot in Belgium and offered new insights on burbot ecology including innovation of captive breeding and the effective reintroduction programme in Flanders.

Thanks as well to both Hendrik Wocher at the LOTAqua hatchery in Germany, and Joachim Claeyé of the AquaLota hatchery in Zele in Belgium, for sharing their knowledge and experience with me.

A number of additional people from around the northern

hemisphere, credited throughout the book, have been kind enough to give me permission to use their photographs of burbot, as well as the types of fishing equipment deployed to catch these fish.

The curious case of the burbot

Burbot are an enigmatic fish in many ways. They are, for example, the only freshwater member of the Gadiformes (the order of cod-like fishes). They are also another of those fishes, like eels, tench and minnows, that look and feel scaleless. This causes prejudice in the eyes of some religious authorities, but in fact they are fully covered by tiny, deeply embedded cycloid scales. Many people find burbot downright ugly, yet they are said to taste delicious. Their liver is not only a delicacy, but also extraordinarily rich in vitamins.

Across their native range, the circumpolar region of the northern hemisphere, burbot are widespread and sometimes common in cool still and flowing fresh and occasionally brackish waters. Indeed, they are amongst the commonest and most widespread fish in Finland, and their populations are also strong across Alaska, Siberia and Mongolia. However, they are under threat in many other parts of their range, often for reasons that are poorly understood but doubtless related to a combination of many factors with habitat loss a principal culprit.

In Britain, burbot became extinct due to a combination of factors. They were also extirpated from Belgium, though now have become established locally due to a reintroduction programme. Their continued existence is also threatened in various other northern European countries and northern states of the United States (US). Some of the common reasons for this decline include habitat degradation, hydrological disturbance, substantial increases in predator populations, construction of dams blocking migration routes

Figure 0.1 Wild burbot on the bed of a lake in the Czech Republic
(© Viktor Vrbovsky)

and compromising feeding, and downstream catchment hydrology and temperature regimes. Breeding and larval life stages appear to be particularly vulnerable, creating bottlenecks to the survival of entire populations. However, as is largely the case in Britain, the reasons for decline are sometimes not definitively known but are likely due to a constellation of multiple factors.

The pursuit of British burbot by angling is necessarily an act of hope, ignorance or blind optimism. Why? Well, the last authenticated burbot to be taken from British waters was landed in 1969. Since then, despite many unsubstantiated claims and rumours, burbot have been conspicuous by their absence from our shores showing up in no fishery surveys.

From a narrowly British perspective, as the fish is absent from our shores at the time of writing, we could ask whether the burbot actually deserves a book of its own. The logical answer from a narrowly British and immediate perspective is a definitive 'No!' However, burbot occupy a broad natural range across which this fish is variously abundant, declining or extinct. Burbot also inspire a variety of interests including inherent, conservation, piscatorial, cultural and gastronomic. This fish most definitely deserves a book of its own; particularly one that breaks through the generalisations about burbot populations, which are oft-repeated and incorrectly applied across the diverse habitats in which regionally adapted strains are found.

Even in the United Kingdom (UK), where distribution was formerly at best patchy, interest in the burbot seems to be undimmed by its long-confirmed absence. Furthermore, at the time of writing, there is a renewed momentum that may shortly lead to the reintroduction of burbot into eastern England.

This is therefore a good time to redress the omission of the burbot from popular wildlife and angling literature; most references in such texts being essentially simplistic copy-and-paste reiterations of accepted wisdoms rather than delving into scientifically based and other realities. After all, the burbot is not only inherently a fish of mystery, but naturally part of our wildlife heritage though extirpated in Britain and other lowland continental European countries within living memory. Burbot are therefore most definitely a fish worthy of the attentions of either the curious or misguided angler and/or naturalist!

Though informed by the most recent and relevant science, I have written this book in an approachable style. References to relevant scientific reviews and other publications are listed in full in the bibliography at the end of the book for those that want to follow up further details. This approach should not, however, disrupt the flow of the book as a good read for those that don't need, or want, to delve into the wider underpinning literature.

I hope you enjoy and find inspiration from this book about an often neglected and frequently poorly treated fish, and share in the hope that, in the coming years, the burbot will prosper once again in the cool waters of eastern England and elsewhere across its broad Palaearctic range.

Chapter 1

Natural history of the burbot

The lineage of the burbot

There are approximately 32,000 living species of fish on planet Earth, a greater number than all of the other vertebrate species (amphibians, reptiles, birds and mammals) combined. Amongst this number, the burbot is unique in being the only 'freshwater cod'.

The burbot (*Lota lota*) is a freshwater member of the Gadiformes, the order of cod-like fishes. (The word 'Gadiformes' literally means 'cod-shaped'.) The Gadiformes are found in marine waters throughout the world, overwhelmingly in temperate or colder regions with a few tropical species that are typically found in deep-water, cooler zones. The Gadiformes comprise ten families of fishes including the Lotidae (the hake and burbot family) to which the burbot belongs. Curiously, this family name is derived from the French word *lotte* meaning 'monkfish', the complete French name for the burbot being *lotte de rivière*.

One of the many oddities of the burbot is that it is the only freshwater member not only of the family Lotidae, but also of the entire order Gadiformes, though a few Gadiformes do enter estuaries. The characteristics of burbot are typical in many ways of that lineage, which puts them at odds with most other freshwater fishes in terms of both their appearance and reproductive habits.

The burbot has an ancient lineage. The oldest fossilised remains of the ancestors of the burbot were discovered in Austria and date back to the Lower Pliocene. (The Pliocene covers the period 5.4–2.4

Figure 1.1 Wild burbot in the bed of Khuvsgul Lake, Mongolia (© Batbold Tsedensodnom)

million years before the present day.) Burbot have weathered ice ages, during which the advance and retreat of glaciers would have wiped out some populations and created new habitats for them (and other cold-adapted fish) to recolonise.

Physical characteristics of the burbot

The body of the burbot is elongated and laterally compressed, almost serpent-like in form, bearing a close resemblance to a catfish or an eel. (For this reason, another common name for the burbot is the eelpout.) The burbot's skin is smooth in appearance and touch. Though seemingly scaleless, the skin is in fact fully covered by tiny, deeply embedded cycloid (smooth-edged) scales much like those of an eel, tench or minnow. The ground colour of the burbot's body is generally yellow, light tan or brown depending on its environment. Their body, head and fins usually have a mottled pattern of dark brown, black or even purple, whilst the belly is yellowish-white. Some fish, noticeably adults from deep lakes in the far north, may be entirely dark brown or black.

The head is flattened dorsoventrally. The eyes are small, reflecting the largely nocturnal habit of the fish and its use of cover in rivers, as well as its adaptation to deeper, darker waters in large lakes. The mouth is large, its rear extent level with the hind edge of the eye, and armed with many small sharp villiform teeth (sharp teeth of equal size resembling close-set fibres). Large, tubular nostrils used for detecting food are located on the upper side of the snout, each with a single, tube-like projection. In common with most other members of the cod-like fishes, the burbot also has a large, single barbel on the chin, described by Wheeler (1969) in his book *The Fishes of the British Isles and North West Europe* as,

> . . . a raised rim on the anterior nostrils appears to form a short barbel.

All of the fins lack spines. There are two dorsal fins. The first, front dorsal fin is short and the second, rear fin is at least six times longer and supported by 67–96 soft rays. To the rear of the underside of

Figure 1.2 Details of the head, eyes, nostrils and mouth of the burbot (© Ian Wellby)

the body, the anal fin is supported by 58–79 soft rays and is almost as long as the second dorsal fin. The second dorsal fin and the anal fin are narrowly separated from the rounded caudal (tail) fin. The pectoral fins are short and rounded. The pelvic fins are narrow with an elongated second ray, and positioned well forward on the body beneath the rear of the gills. Whilst the pelvic fins are pale in colour, the other fins are dark and mottled. The small size of the fins relative to the body is indicative of the benthic lifestyle of this fish, adapted to life in still or slow-flowing water. Although burbot are generally considered to be unable to withstand strong currents, they are nevertheless found in often quite torrential rivers also inhabited by fish such as trout. Riverine burbot populations can also undertake prodigious spawning migrations. For example, one Passive Integrated Transponder (PIT) tagged German burbot was recorded migrating to a spawning area 254 km upstream of its summertime location.

Within the body of the fish is a gas-filled swim bladder, enabling the fish to adjust its buoyancy. Burbot can also use their swim bladders to make sounds, which may be important as a means of communication during spawning.

Normally, burbot do not live beyond 15 years, but ages as high as 20 years have been recorded. They can reach a maximum body length of around 1.5 m (about 5 ft), with a maximum published weight of 34 kg (nearly 75 lb). However, burbot in river populations are generally much smaller. Maximum size varies with habitat occupied, geographical region and the regional genetic strains found there. As we will see when addressing other aspects of burbot ecology, generalisations derived from studies of populations from large lakes in the United States (US) and Scandinavia may not automatically apply to riverine populations. By and large, the genetic strains of burbot found in European rivers, Western European populations in particular, tend to be much smaller than their large-lake cousins.

Burbot habitats and habits

Though some generalisations can be made about burbot, we have to bear in mind that they have a very broad northern hemisphere distribution and therefore make use of a wide range of habitats, as Nick Giles (1994) comments in his book *Freshwater Fish of the British Isles: A Guide for Anglers and Naturalists,*

> In its widely differing habitats across a wide geographical range the burbot lives with trout and grayling in rivers or with a range of cyprinids at the northern edge of their distribution in cool lakes.

Adult burbot prefer cold conditions, occurring in deep water in lakes and slow-flowing rivers. However, burbot also inhabit the smaller rivers of northern Europe, within which are found adapted strains. These smaller fishes tend to seek refuge in tree roots, rock and woody debris and other structures, and can occur in some quite fast rivers.

Although burbot favour fresh water, they are also found in some brackish environments. They are particularly common in the low-salinity margins of the Baltic Sea, from the coasts of eastern Denmark and Germany through Poland, Lithuania, Latvia, Estonia, Russia,

Figure 1.3 Wild, fast river habitat occupied by burbot in southern Germany (© Hendrik Wocher)

Figure 1.4 Tree roots are favoured burbot habitat in riverbanks
(© Johan Auwerx/INBO)

Finland and Sweden. Here they can be found in an unfamiliar asso-ciation with a mix of marine and freshwater species that may include cod, pike, herrings, roach and ide. They are particularly common around the coast of Finland, especially in the Gulf of Bothnia and the Gulf of Finland where, due to inflows of major river systems, the salinity is very low.

Burbot are benthic fish, meaning that they live on the bed of the water body, though also inhabit crevices in riverbanks including lurking in tree roots. They tolerate a diverse range of substrate types from mud to sand, rubble, boulder, silt and gravel. Adult burbot may construct extensive burrows in soft sediment or make use of natural caves or cover, in which they shelter by day. They emerge in the hours of darkness, with peak activity at dawn and dusk, pursuing an entirely predatory lifestyle.

In large lakes, burbot generally retreat to deeper, cooler and less accessible waters during summer months, only coming into shal-lower water to spawn in mid-winter. For this reason, burbot may be rarely encountered in the accessible margins of larger and deeper lakes for much of the year. Burbot can live at great depth, having been recorded at a depth of 300 m (980 ft) in Lake Superior, North America. They can also occur at altitudes up to 1200 m in the cooler waters of alpine lakes. These deeper waters can become deoxygenated as still waters get covered in ice in winter, leading the burbot to move to shallower water in search of food and more richly oxygenated water. Spawning may occur under the ice or in the shallower waters of lake margins in winter.

In prealpine lakes, burbot larvae are pelagic (living in open surface water), moving with, and feeding on, zooplankton. However, they move to the littoral zone (around the shores) after metamorphosis, adopting a benthic lifestyle thereafter. After the first summer, most burbot fingerlings migrate into the profundal zone (deep water) where they live out most of their adult lives on the lakebed.

In rivers, the favoured habitats of adult burbot include undercut banks beneath large rocks and woody debris, and in particular amongst tree roots. Adult fish also seek out the deeper water of main

Figure 1.5 Adult burbot are benthic, lake specimens living out their adult lives in cover in deeper water (© Batbold Tsedensodnom)

channels of larger rivers for much of the year. The presence of larval and fingerling burbot almost exclusively in smaller tributaries and still or slow-flowing pools, as well as other wetlands on the flood-plain, suggests that these river-dwelling burbot migrate upstream into shallower water and riparian habitats within river systems to spawn. The presence of extensive, connected riparian wetlands is particularly important for spawning and nursery purposes. Shading by trees may be important both for the diversity of habitats that they provide, including underwater root systems, as well as helping reduce excessive river temperature through shading.

In both rivers and lakes, experiments with tagged fish confirm more general observations that burbot show high 'site fidelity' (philopatry), remaining in the same location in deep water for the whole summer and only becoming mobile as they head inshore or upstream to spawn in the winter. They may also return to the same haunts after spawning.

In habit, burbot are predominantly nocturnal exhibiting peak activity around dusk and dawn. In deep lakes and larger rivers with slow-moving currents burbot tend to remain in deeper water, whilst in rivers they typically secrete themselves in refuges, often sheltering

Figure 1.6 Groups of burbot may be found sharing a suitable habitat, such as this shipwreck in Khuvsgul Lake, Mongolia (© Batbold Tsedensodnom)

under rocks, in crevices under riverbanks, among roots of trees and dense vegetation. In both still and flowing waters, burbot of similar sizes may form communal groups sharing the same favourable habitat features. As darkness falls, burbot tend to emerge to move into shallower water in search of food.

The burbot diet

Although sometimes described as omnivorous, burbot are in fact strongly predatory and can also exhibit cannibalistic tendencies. Smaller burbot feed on a varied diet largely comprising invertebrates such as worms, molluscs, insect larvae and crustaceans. However, larger burbot show a distinct preference for eating other fish and, as top predators, are sometimes considered 'the lion of underwater society'. The whitefish (coregonid) populations of some of the larger, cooler lakes where burbot occur often form an important part of the burbot diet. As discussed later in this book (see Everard, Chapter 5, this volume, 2021), the burbot may have a role in controlling some problematic alien invasive fish and other animal species.

Smaller burbot are preyed upon by many fishes including pike, zander, walleye (American zander), wels catfish and eels. As younger burbot in suitable habitats may be present in early spring in massive numbers, they represent an important food source for the early life stages of these more desirable fish species. However, larger burbot also prey upon the young of these more commercially valued species, which may make the fish unpopular with commercial and some recreational fishermen.

A study of the stomach contents of 5253 burbot captured near various ports along the north shore of Lake Erie in 1946–1947 found that their diet varied in relation to method of capture, season, locality, and the length of the burbot (Clemens, 1951). Fish and invertebrates, principally crustaceans, constituted a major part of the diet of young burbot. After their first summer, when they may have attained a length of 10–15 cm, these lake burbot were observed to shift their diet from invertebrates to small fish that thereafter comprised the predominant part of their diet. Perhaps unsurprisingly, larger burbot were found with a greater volume of stomach contents, and tended to consume larger fish, rather than larger numbers of smaller fish. More food was found in the stomach contents of burbot in winter compared to the summer, possibly reflecting a lower rate of digestion in colder water. The study also concluded that burbot had to compete for food with other carnivorous fish species during all life stages.

Reproduction and growth

Burbot reach sexual maturity at an age of between two and seven years, depending on local conditions, across much of the cool, pan-Arctic climatic range in which they exist. At the Linkebeek fish conservation hatchery operated by the Flemish government's *Instituut voor Natuur-en Bosonderzoek* (INBO) in Belgium (Box 1.1), captive burbot spawned from broodstock sourced from European rivers were found to mature at around two years of age, and exceptionally one year, once they reached a length of around 18 cm (7 in).

Spawning occurs in the winter, consistent with that of other marine cod-like fishes. With lake populations, this can often occur under ice at extremely low temperatures between 1°C and 4°C, and generally at night when temperatures are coolest. However, the optimal spawning temperature and the sensitivity of eggs and larvae appears to differ for lowland river populations, with successful spawning of burbot from river populations achieved in captivity at 5°C. Paragamian and Wakkinen (2008) suggested that temperatures below 6°C may be an important trigger for spawning migrations by adult burbot. These observations were supported by Żarski et al. (2010) who found that spawning could not be induced in captivity in Poland at temperatures greater than 6°C.

Burbot vocalise during the spawning period. They produce a range of sounds by rapidly contracting the drumming muscles connected with their swim bladders, akin to those found in marine cod-fishes. Underwater acoustic monitoring experiments in an under-ice enclosure in Canada found that these low frequency sounds were most frequent during the burbot's spawning period (Cott et al.,

> ## Box 1.1 'Instituut voor Natuur-en Bosonderzoek (INBO), Belgium'
>
> INBO is the *Instituut voor Natuur–en Bosonderzoek* (the Research Institute for Nature and Forest), an institution of the Flemish government established to underpin and evaluate biodiversity policy and management through applied scientific research, data and knowledge access. INBO also provides information for international reporting, and supports nature management queries from local authorities and other related organisations.
>
> Much of the work of INBO is commissioned by the *'Agentschap voor Natuur-en Bos'*: Agency for Nature and Forest (ANB) of the Flemish government. ANB initiated recovery programmes for the reintroduction or support for declining aquatic species, issuing instructions to INBO to breed different species for reintroduction and to undertake subsequent monitoring. INBO is also tasked with carrying out and implementing scientific research to deliver this reintroduction programme.
>
> INBO operates a Center for Fish Farming, based in Linkebeek in the Belgian province of Flemish Brabant just south of the capital, Brussels. Burbot and other species are bred for conservation purposes at this hatchery site, which is also a highly active and productive centre for related research.

2014). Recorded acoustic data revealed a wide repertoire of calls ranging from slow knocks to fast buzzing, similar to those produced by haddock (*Melanogrammus aeglefinus*) and other cod-like fishes. Vocalisations are known to be important in the mating systems of other codfish species. It is assumed that calling by fish under ice cover, or in complex wetland habitats, is likely to be an important part of the mating system of burbot and other spawning codfishes, as reduced light levels under ice or at night are likely to reduce the usefulness of visual cues. The extent to which vocal communication is affected by increasing levels of underwater noise generated by human activities is unknown.

Spawning may occur as early as December, and as late as mid-April, across the broad northern hemisphere circumpolar range of the burbot. The spawning season is short in any locality, lasting only two to four weeks. During this short window of time, the female burbot spawn only once but males may do so multiple times. However, spawning may not occur every year.

For lake burbot, spawning generally occurs in the shallow margins of the water bodies in which they occur. Mature burbot tend to migrate into shallower water for spawning with male fish usually arriving first and females following a few days later. Spawning usually takes place at depths of less than 3 m (about 9 ft), with favoured habitats including sandy or gravel beds near the shore or else on mid-channel shoals.

In river systems burbot are adfluvial, meaning that they spawn in tributary streams and their connected wetlands. The larger rivers in which they spend much of their adult lives generally lack the kind of still or sluggish margins necessary for the survival of eggs and juveniles. Consequently, burbot undertake often extensive spawning migrations – distances of 100 to 200 km are not unknown – to access headwaters or areas of floodplains with a mosaic of standing or slow-flowing water. These areas provide rich zooplanktonic food (small animals such as rotifers and cladocerans, or 'water fleas', suspended in the water column) and allow the hatching of eggs and subsequent growth of hatchlings to occur without risk of wash-out. The availability of such diverse wetland habitats in the floodplains of river systems, and their inundation over a period of months without rapid hydrological change, can be a critical bottleneck for burbot recruitment and hence the survival or revival of riverine populations.

Burbot are broadcast spawners, meaning that they do not have an explicit nesting site or medium. The act of spawning is communal and nocturnal. Many male burbot can gather around one or two females, forming a 'spawning ball' comprising as many as a dozen fish clustering into tight balls in open water.

Burbot have high fecundity. Female fish, which mature typically after two or more years, release anything from 63,000 to 3,478,000 eggs depending on body size. Experimental observation at the Linkebeek fish conservation hatchery operated by INBO in Belgium found that relative fecundity of broodstock burbot is 663,000 eggs per kg body mass. The eggs are tiny, each with a diameter of around

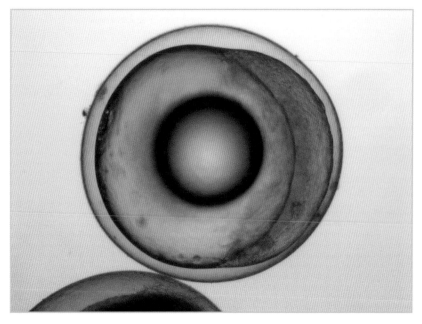

Figure 1.7 Burbot egg 48 hours after fertilisation (© Ian Wellby)

1 mm (about the size of a pinhead), and are generally coloured light green, though they may take on a pale orange colour when the parent fish have been feeding on freshwater shrimps or other prey with strong carotenoid pigments. Sperm from the attendant male fish is released simultaneously with the release of eggs by the female. Fertilised eggs drift in the water column after release, slowly coming to rest on the bed of the lake or river where they may get caught in voids or cracks in gravel or other substrates.

The temperature at which burbot eggs are incubated is critical for their survival. Experiments at INBO's Linkebeek fish conservation hatchery, with burbot broodstock derived from Western European lowland rivers, found that optimal egg survival occurred at a temperature of 3.8°C, with 100% egg mortality recorded at temperatures exceeding 8.1°C (Vught et al., 2008). However, this low temperature is not essential throughout the entire egg incubation period. The early days were found to be most critical. The water must be cold enough at the time of egg deposition as burbot sperm becomes immobile above 6°C. A temperature around 4°C is

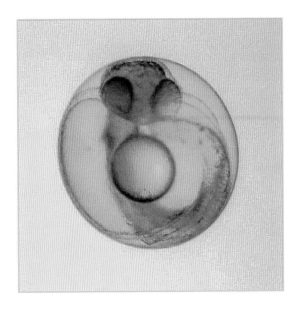

Figure 1.8 A 1 mm burbot embryo just before hatching (© Johan Auwerx/ INBO)

important during the first ten days of incubation, after which eggs can withstand higher water temperatures. Overall, a period of ten to fourteen days over which the water temperature drops to 4°C is required to enable successful reproduction. Some other studies have found total egg mortality at temperatures above 5°C or 6°C, though this may reflect greater sensitivity between different genetic strains of burbot.

Like all cod-like fishes, burbot exhibit no parental care. The eggs are instead left unattended, at the mercy of currents and a wide range of predators, with a correspondingly high mortality rate. The adult fish return to the deeper water of lakes or marginal habitats in the main channels of lower rivers after spawning. Tracking experiments in the US and Denmark suggest that adult burbot are loyal to their habitats, returning to the same holes after spawning.

Burbot egg incubation is typically 120–135°C days (a multiplication of temperature in degrees Celsius by the number of days). This generally amounts to 30–128 days depending on ambient water temperature. Consistent with the small egg size, the immobile hatchlings (still attached to a small but visible near spherical yolk sac) are also tiny. Lake populations tend to produce larvae that are barely 4–4.5 mm in length. Hatchings of the smaller genetic strain of burbot derived from smaller Western European rivers were found in INBO's Linkebeek fish conservation hatchery to be 3–3.7 mm in length. During this period, the larvae have no gills, no mouth, no developed fins, and are totally helpless.

The yolk sac of the burbot differs from that of other freshwater fishes in both its spherical shape and its small size at around 17% of total larval body length, compared for example to the elongated yolk sacs of dace and chub comprising respectively about 47% and 60% of larval body length. The yolk sac of the burbot larva is also characterized by the presence of a large oil droplet, occupying about 30% of the entire yolk sac at hatching. This oil droplet is absorbed more slowly than the yolk and persists for some days after the yolk is absorbed, suggesting that it has an important secondary function in increasing the buoyancy and orientation of the larvae before the swim bladder inflates (Palińska-Żarska et al., 2014). Currents can distribute the larvae over large distances during this immobile, buoyant phase. The hatchings remain inert whilst they absorb this

Figure 1.9 Freshly hatched burbot larvae, 3 mm long, at the INBO fish conservation hatchery (© Johan Auwerx/INBO)

yolk over a period of between two and four weeks, duration again largely related to water temperature.

Temperature remains critical for ongoing development as burbot larvae are intolerant of higher water temperatures. Experimental breeding of lake populations of burbot in Idaho has found that embryos spawned at 2°C had a survival rate of 86.7% to the eyed stage, but this declined to 47.9% and 0.1% if spawning and incubation occurred at 4°C and 6°C respectively, with embryo deformity increasing dramatically between 4°C and 6°C (Ashton et al., 2019). The authors of this study concluded that spawning temperatures above 4°C potentially underlie burbot recruitment bottlenecks in systems affected by impoundments, climate warming or other barriers to cold-water spawning habitat. However, as observed in terms of optimal survival of eggs, there are large differences between the various genetic strains of burbot across their wide geographical range, with significantly higher temperatures tolerated by the larvae of burbot from Western European lowland rivers (Vught et al., 2008). This highlights the importance of crude generalisations, and also of selecting appropriate genetic strains in any fish reintroduction or support programmes.

Observations of larval burbot in experimental conditions found that inflation of the swim bladder begins from the third day following hatching, the yolk and particularly the oil droplet playing roles in buoyancy until this time. Half of the larvae commenced feeding after the ninth day, and complete yolk absorption occurred after two weeks (Palinska-Zarska et al., 2014). The transition to external feeding is critical for the survival of larval burbot, with an associated high mortality rate.

Larval burbot then migrate towards the surface as part of the plankton, much like other marine cod-like species. In colder lakes, this can occur around May, though the end of March to the beginning of April is more common for lowland river burbot. Here, the larval fish feed on the spring bloom of small planktonic animals. The availability of sufficient small zooplanktonic food is critical for larval survival, with food needs varying by genetic strain and larval size. Research conducted in a lake in north-eastern Poland found that larval burbot fed extensively on copepods (teardrop-shaped water

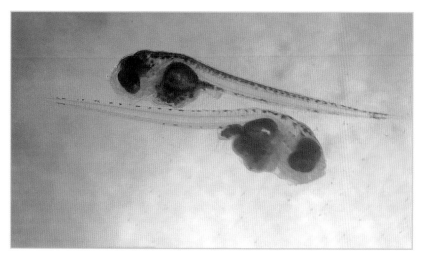

Figure 1.10 Burbot larvae starting exogenous feeding
(© Joachim Claeyé/AquaLota)

fleas typically 1–2 mm (0.04–0.08 in) in length with large antennae), moving on to selectively consume cladocerans (larger, nearly
spherical water fleas ranging from 0.2–6.0 mm (0.01–0.24 in) in
length) as the fish grow (Furgala-Selezniow et al., 2014). Whilst
these Polish burbot larvae were found not to eat planktonic rotifers
(microscopic 'wheel animalcules') even where they were numerous,
apparently due to the small size of these planktonic invertebrates,
rotifers were critical first food for the larvae of the Western European
strain burbot that are initially too small to ingest copepods. During
the larval stage, burbot grow from a length of around 4.0 mm to
20–30 mm. Throughout the warmer months, juveniles begin to
adapt to a progressively more active lifestyle whilst continuing to
feed largely on aquatic invertebrates.

Burbot larvae then metamorphose into juveniles that drop to the
bed to adopt a subsequently entirely benthic lifestyle. From this stage
onwards, they feed on benthic invertebrates until, attaining a length
of around 20 cm (8 in), they consume an increasing proportion of
other fish (including other smaller burbot) though the diet is largely
opportunistic. Though juvenile burbot may take shelter during the
day under rocks and other debris, they are active by night. Burbot
derived from lowland Western European river broodstock in the

fingerling stage were found to have thermal tolerance up to 20°C, with best growth rate achieved around 16°C (Harzevili et al., 2004). Thermal tolerance, particularly in the egg and early larval stage but also through juvenile life stages, may impose a significant limitation on their survival and spread, though is variable with genetic strains across global burbot populations.

Juvenile burbot grow rapidly during their first year. Wheeler (1969) states in *The Fishes of the British Isles and North West Europe*,

> One year old burbot measure 6–6¼ in (15–16 cm) on average, two year old fish 8½–9 in (22–23 cm), and at maturity in their third year, usually 12¾–13½ in (32–34 cm).

Experts in burbot hatcheries say that juvenile burbot typically attain a length of 11–12 cm (approximately 4 in) by the start of the following winter, although some individuals grow markedly quicker than others with lengths up to 25 cm (approximately 10 in) after the first year not unusual. The cannibalistic habits of these larger individuals can present problems in aquaculture.

In river systems, larval and fingerling burbot are found almost exclusively in small tributaries. This is strongly indicative of the

Figure 1.11 Eight month old burbot, mainly around 10 cm but with some cannibalistic specimens up to 25 cm (© Johan Auwerx/INBO)

importance of marginal wetland habitat with static water or slow flows for successful reproduction. It is essential that wetlands connected to the river system contain water from December onwards to enable adult burbot to migrate to their spawning grounds, as well as for the development of immobile eggs and larvae with yolk sacs attached in January and February, and to allow the young larvae to locate zooplanktonic food in March and April.

In large lake populations in North America, burbot reach adulthood at an age of around three to five years at a body length of around 40 cm (16 in) maintaining the benthic lifestyle adopted since metamorphosis. At this transitional stage, they migrate to the bed of deeper lakes and into bigger and deeper rivers to live out their adult lives, feeding predominantly on fish and other animals.

By contrast, burbot strains found in Western European lowland rivers are smaller, making use of available cover such as under rocks and other obstructions as well as tree roots and woody debris. They also typically mature after two years, attaining smaller body lengths.

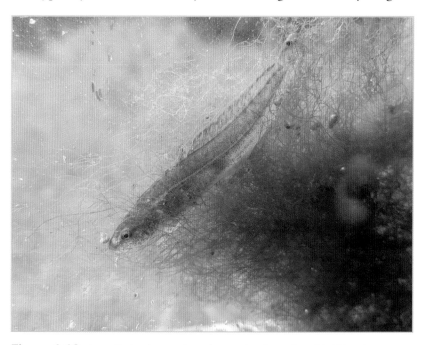

Figure 1.12 Juvenile burbot at 6 cm long adopting a benthic lifestyle (© Joachim Claeyé/AquaLota)

Figure 1.13 Wild burbot using rock cover in the Barkh river, in the Pacific drainage of Mongolia (© Batbold Tsedensodnom)

Other river populations, such as the burbot in the rivers of the Pacific drainage in Mongolia, are also smaller than those occupying lakes.

Dependent upon local climate and genetic strain, burbot mature after between two and seven years to repeat the cycle.

Predators of burbot

Although burbot are voracious predators of other fish, they are also consumed by other larger fish species. Burbot larvae, as we have seen, are tiny and may be numerous, and are preyed upon by a diversity of fish, invertebrates and other organisms. As the larvae metamorphose and adopt a benthic lifestyle, they seek cover but are far from immune from predation by larger fishes including notoriously cannibalistic larger burbot. Whilst still occupying shallower water, predatory birds such as herons, egrets, kingfishers and grebes may predate upon juvenile burbot, as may otters and other predatory aquatic mammals.

As burbot grow and head into deeper water, bigger pike, zander, wels catfish, eels and other predators may eat them when they venture out of cover to feed. Giles (1994) comments in *Freshwater Fish of the British Isles: A Guide for Anglers and Naturalists* that,

> German studies record eel diets including sticklebacks, burbot and fish spawn.

Burbot genetics

As indicated when considering habitat and reproductive traits, burbot display significant genetic diversity across their broad geographical range. The collective results of research studies by Van Houdt et al. (2003 and 2005) suggest that there may be two sub-species: *Lota lota lota* distributed throughout the Eurasian range including Alaska in the US (formerly connected by a land bridge to Asia); and *Lota lota maculosa* in North America south of the Great Slave Lake in Canada. (Great Slave Lake is the second-largest lake in the Northwest Territories of Canada, the deepest lake in North America and the ninth-largest lake in the world.) This division is considered to have resulted from climatic oscillations during the Pleistocene epoch, with its consequences for glacial refugia and post-glacial colonisation routes.

There are smaller genetic differences between populations in different drainage basins in North America. In their recent work in and around eastern Lake Michigan and Michigan tributaries, Blumstein et al. (2018) found a 'moderately high' genetic (microsatellite loci) diversity between samples collected from Lake Michigan and tributaries of the Manistee River. These results indicate reproductive isolation between riverine and lacustrine spawning life history types that, as noted above, have different migration patterns. This is also consistent with the high degree of philopatry (tendency of an organism to stay in or habitually return to a particular area, including returning to similar spawning locations) observed in burbot. In Europe, Wetjen et al. (2020) found distinct genetic differences in the mitochondrial deoxyribonucleic acid (DNA) of burbot between major river basins in Germany, and particularly between lake and river habitats.

Analysis of burbot DNA across Europe has identified four distinct phylogenetic clades (groups of organisms believed to be evolutionary descendants of a common ancestor). These are identified as Northern European (Scandinavian), Central European, Baltic (a mixed population comprising features of the Northern and Central clades) and Western European (Van Houdt et al., 2003 and 2005). Research by INBO found that French burbot populations, part of the Western European clade, were found to be closest genetically to the original stock that had since become extinct from Belgian rivers. Furthermore, habitat in the La Bar river in France was observed to be similar to that of Flemish lowland rivers, suitable for the needs of the different life stages of the burbot. Consequently, wild-caught French burbot were used as broodstock for reintroduction to Belgium, and the habitat occupied by burbot in French rivers helped identify optimal potential reintroduction sites in Belgium. The Belgian burbot reintroduction programme is addressed later in this book (see Everard, Chapter 5, this volume, 2021).

Samples collected from 15 museum specimens of burbot of known English provenance were analysed for differences in mitochondrial DNA (Worthington et al., 2010a). This genetic analysis suggested that the extinct English burbot population was a distinct lineage from those previously described from across the global distribution, but that it was most closely related to the Western European genetic strain. This supports the belief that burbot colonised Britain via the Doggerland land bridge formerly connecting these islands with present-day continental Europe (see Everard, Chapter 2, this volume, 2021), surviving the peak of the ice age in freshwater refugia. This land bridge explains the former range of burbot, as well as that of many other freshwater fishes, across Britain.

Although genetic studies have not been carried out extensively east of the European range, geographical barriers can reasonably be expected to have given rise to differing genetic strains of burbot adapted to local aquatic conditions. As one example, Mongolia is divided into three main basins: the Arctic, Pacific and the landlocked Central Asian. Streams originating from a mountain towards the centre of the country form headwaters of these diverse catchments, fanning thousands of km in different directions into landscapes of

Figure 1.14 Cased burbot specimen from 'the River Trent near Nottingham' dated March 1905 (institutional abbreviation 'NOTNH V2732P' from the Nottingham Natural History Museum, Wollaton Hall. DNA was extracted from museum specimens to determine the genetic profile of extinct English burbot (© Nottingham City Council/Nottingham Natural History Museum, Wollaton Hall)

differing and extreme climates. Substantial genetic radiation can reasonably be expected, adapting localised strains of burbot to differing river, lake and other wetland habitats and climatic regimes across Mongolia. Burbot from riverine populations in Mongolia are visibly smaller than their lake-dwelling relatives.

This fine-scaled differentiation of genetic make-up within apparently similar local burbot stocks may have significant implications for the development of effective management options to preserve genetic diversity and particularly for natural recruitment. It has implications for future stocking programmes and the preservation of the adaptive potential of burbot, including potential burbot reintroductions.

Chapter 2

Burbot distribution

Burbot have a broad, northern circumpolar distribution including both Eurasia and North America, generally north of the 40° latitude. They have the second-broadest natural distribution of any larger freshwater fish, after northern pike (*Esox lucius*). This distribution is limited by the burbot's preference for clean, well-oxygenated water. Across much of this range, this may be in water temperatures of less than 13°C. However, lowland river populations can withstand substantially higher temperatures in mid-summer. Genetic evidence suggests that this cold tolerance has enabled burbot to rapidly colonise new fresh waters as glaciers retreated after the last ice age, spanning these cooler regions of the northern hemisphere.

Threats and trends across their natural range

Burbot were formerly widespread and abundant throughout much of their natural range, though have not thrived in all regions over

Figure 2.1
Approximate Holarctic distribution of burbot, adapted from Stapanian et al. (2010) (© Dr Mark Everard)

the past centuries. In fact, they have declined to eventual extinction in Britain and some other countries throughout the 20th century. A review of burbot captures from the early 19th century up to 1970 by Marlborough (1970) concluded that,

> Local over-fishing, pollution and habitat changes are considered the most likely causes of decline. Conservation measures seem desirable.

A review by Stapanian et al. (2010) concluded that many populations have been extirpated, were endangered or were in serious decline by 2010. Factors suggested by the review's authors as being responsible for this decline included pollution and habitat change, particularly the effects of dams affecting riverine burbot populations. The adverse effects of introduced invasive species was seen as being one of the main reasons for declines in lake populations. Warmer water temperatures, due either to discharge from dams or climate change, were also noted as potential contributors to declining burbot populations at the southern extent of their range.

Fishing pressure did not appear to be limiting burbot populations worldwide. However, there are possibly some circumstances where fishing pressure may be significant, for example, locally in Wyoming in the United States (US) at the southern boundary of the burbot's range. Issues of overexploitation, entrainment in irrigation canals and habitat loss were all suggested as causes of the decline of burbot in a 2017 study that explored the potential role of overexploitation through capture–recapture experiments with tagged fish in a range of US lakes (Lewandoski et al., 2017). It was concluded that there was a low risk of overfishing in two lakes, intermediate risk in one lake and high risk in another. This study cannot be assumed as representative of burbot populations more widely, as it is at the south of the range and burbot populations may possibly be under further stresses.

Temperature regime at critical times of the year appears to be particularly influential on the viability of burbot populations. Based on substantial experience since around 2005, Hendrik Wocher, who operates the LOTAqua hatchery in Germany (see Everard, Chapter 5, this volume, 2021), observes that burbot are tolerant of summer peak water temperatures of over 20°C. The Oder river, forming part of the border between Poland and Germany, peaks at 24–25°C in summer during which time larger burbot migrate to the estuary with the Baltic Sea, whilst smaller fish tend to migrate upstream to small tributary rivers. Here, burbot can remain inactive for long periods of time subsisting on reserves stored in their very large livers (comprising as much as 15% of body weight of fish in aquaculture and 8–10% in wild fish). However, tributaries of the Oder have a maximum temperature of 4°C in winter. This short-term toleration of higher temperatures is endorsed by findings reported by Johan Auwerx at the *Instituut voor Natuur-en Bosonderzoek* (INBO), also observing that larval and juvenile burbot grown on in rearing ponds could withstand peak summer temperatures of up to 30°C, becoming torpid and ceasing to feed.

A period of four to six weeks at 4°C or below has been observed in US large lake burbot populations as being essential for maturation of adult fish, breeding and early larval development, though higher temperature tolerances are observed in lowland Western European populations. There is therefore no certainty that burbot can successfully breed every year in every location. However, their longevity, including maturation at an age as early as two years, as well as the high fecundity strategy may mean that successful spawning every few years when conditions are ideal may be sufficient for population viability.

Of all the factors implicated in the demise of the burbot, habitat loss is perhaps the most significant for river populations. As will be described in greater detail throughout this book, river straightening, land drainage and disconnection of channels from their formerly extensive and diverse floodplains robbed rivers of the mosaics of wetland habitats essential for burbot spawning, survival of free-floating eggs and largely immobile larvae, and plankton-rich nursery areas for juveniles. Owing to this extreme vulnerability to habitat loss, burbot have been considered 'the ambassadors of the wetlands'.

Also, as vital breeding and nursery habitat for many species was eradicated for farming, inputs of fertiliser, pesticides and other chemicals degraded the quality of remaining water bodies. Up to the year 2000, half of all the nitrogen fertiliser ever applied globally had been spread on fields since just 1985 (Millennium Ecosystem Assessment, 2005) whilst, in the 50 years preceding 2013, overall global fertiliser use had increased by about 500% (Juniper, 2013). In fact, a rich cocktail of man-made chemicals enters river systems from farming operations, pet-care products (such as flea treatments for dogs) and a range of household and gardening products, all of which can affect the quality of the zooplankton, and in turn compromise the survival of larval and juvenile burbot and the health of the wider ecosystem.

All of these factors have combined to result in often significant threats to burbot populations, asymmetrically so in river populations, across their natural range.

Burbot distribution in North America

In North America, burbot are widespread from the Great Lakes northwards into Canada, and into adjacent states of the US. Burbot are particularly well distributed throughout the colder regions of Canada, aside from the extreme west of British Columbia, the eastern province of Nova Scotia and the Atlantic Islands.

They are also widely distributed across the North America state of Alaska. Across Interior Alaska, burbot are abundant in large glacial river systems such as the Tanana, Yukon, Kuskokwim and Copper Rivers, and they also reside throughout the year in small, clearwater tributaries.

Burbot distribution in Eurasia

In Eurasia, burbot occur in clean lakes, rivers and some brackish waters in 19 countries across the northern continent. However, they are absent from the Iberian Peninsula and Greece. They are also absent from the Kamchatka Peninsula, a 1250 km (777 mile) long peninsula in the Russian far east, as well as from the west coast of Norway.

Burbot naturally occur across Siberia, Mongolia and elsewhere in Northern Asia as far as the 45th parallel to the south. However, in Europe, burbot have spread further south than this approximate boundary in some rivers, including the Rhône, Seine, Loire, Danube and the lower reaches of the Kura.

A review by Stapanian et al. (2010) concluded that burbot had been extirpated or were in decline in much of western Europe and had been lost entirely in the United Kingdom (UK). Across the Netherlands, Germany and Belgium, burbot were found to be extensively imperilled, vulnerable or extinct, though targeted conservation programmes were in place in some of these countries to address that trend. Restoration activities in the Lippe catchment in Germany are considered in more detail (see Everard, Chapter 5, this volume, 2021), as is the successful burbot reintroduction programme in Belgium where the species had been extirpated. A study of historic data by Bosveld et al. (2015) found that the distribution and population trends of burbot in the Netherlands were relatively stable at the beginning of the 20th century, though entered a declining trend from around 1950. This date is consistent with declines in many other species across Europe, coincident with major intensification of agriculture. Only two areas of the Netherlands were found to contain spawning populations at the time of the 2015 study – lakes in the confluence area of the rivers Vecht and Ijssel – where an observed recent increase in burbot numbers was attributed to annual stockings in German reaches of the Vecht river since 2001. However, a third Dutch population has been established in the De Beerze river near the village of Oirschot as a result of a reintroduction programme using juveniles from INBO's Linkebeek fish conservation hatchery between 2009 and 2013. This was done in conjunction with a floodplain restoration programme resulting in inundated wetland habitat throughout the winter from which juveniles resulting from natural reproduction are now caught in surveys.

Burbot were also found to be threatened or endangered in much of the remainder of Eurasia, though the Siberian population seems to remain strong.

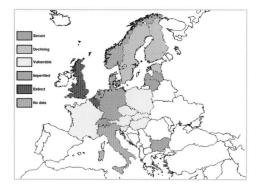

Figure 2.2 Assessment of the vulnerability of burbot populations by European country, adapted from Stapanian et al. (2010) (© Dr Mark Everard)

Burbot distribution in Britain

In Britain, burbot were native to eastern England only. They are naturally absent from Ireland and Scotland as well as Wales, as well as the west and much of the north of England.

Their former English distribution is substantially explained by the land bridge, known as Doggerland, formerly connecting modern-day Britain as a peninsula of the continental European land mass. During this time, the Rivers Thames, Rhine and Scheldt formed the Channel River, which carried their combined flow to the Atlantic. Towards the end of the last ice age (roughly 6500–6200 BCE), a massive ice lake to the north breached releasing a megatsunami. This catastrophic event inundated Doggerland, now an underwater relic known as the Dogger Bank in the North Sea, severing the connection between the British Isles and continental Europe. The Channel River was no more, now constituting the English Channel, and the formerly connected river systems were separated into discrete drainage basins.

Many freshwater fish species lacking life stages tolerant of saline water (the term 'stenohaline' is used to describe organisms that cannot tolerate a wide fluctuation in the salinity of water) – barbel (*Barbus barbus*), gudgeon (*Gobio gobio*), and many more – consequently had a natural distribution only in the river catchments of eastern England that had formerly been connected to the Channel River system. Access to other river systems to the west and north was effectively prevented by impenetrable dry land barriers.

Burbot were therefore formerly resident in eastern rivers of England only, where distribution may have been patchy due to their specific habitat needs. Historic texts range from descriptions of burbot being so abundant in the 16th century that they were scooped out to feed pigs or, alternatively, that they were already in decline by the 19th century. British burbot subsequently went into precipitous decline and eventual extinction.

The burbot is one of the last vertebrate animals to go extinct in England. They did so during the middle of the 20th century, at the same time that many other species, including salmon and sea trout, also slumped and were lost in eastern England. Many of the eastern rivers that were their prior strongholds were not strongly industrial, suggesting that mass industrialisation in the preceding period was only one of many likely contributory factors.

Agricultural intensification and widespread land drainage were far more strongly implicated in the eradication of the complex of habitats required by river-dwelling burbot, particularly of early life stages. Massive investment in river straightening and land drainage for agricultural intensification as a contribution to food security following the Second World War is a highly likely major culprit in the final extirpation of burbot, habitat degradation and fragmentation compounded by profound influences on riverine chemistry and

Figure 2.3 Straightened channels separated from their floodplains by embankments, like the Bevill's Leam in the Fenland district of Cambridgeshire (England), offer no functional habitat for the breeding and nursery needs of burbot (© Dr Mark Everard)

hydrology. There are similar recollections of the precipitous decline of the once common burbot from elsewhere in lowland continental Europe, where river straightening in the 1950s also prevented sufficiently long-term inundation of floodplains. Some river locations in France where burbot populations remain strong today have had no such history of major conversion for agriculture.

In fact, the history of land drainage across the east of England, as also elsewhere across the European range where the burbot populations have declined or become extinct, have a far longer history. Post-war drainage has simply been a final nail in the metaphorical coffin. As Mark Cocker (2018) describes in *Our place: can we save Britain's wildlife before it is too late?*,

> The wetland known as the Fens, of which Gedney is a part, once covered 1500 square miles from the outskirts of Cambridge in the south, as far as Peterborough in the west, and almost to Lincoln and Boston to the north. In Roman times, it was an immense semi-circular expanse of mire, mere and low-lying marsh, with the maritime bay known as The Wash at its heart.

The long history of progressive drainage of the Fens dates back at least to Roman earthworks and dykes. These were progressed by Saxon monks, amongst others. The most famous of subsequent drainage projects were initiated in the years before the English Civil War south of Downham Market and west of Ely in an area formerly known as the Great Level. This was the major earthwork and channelisation programme for which Dutch engineer Sir Cornelius Vermuyden was commissioned, and whose name is still associated with various structures and places on the levels. Vermuyden was already something of an agricultural improvement celebrity across Europe, knighted in Britain for earlier works seeking to drain formerly extensive wetlands on the Yorkshire–Lincolnshire border. Vermuyden's programme of drainage of the Fens spanned two phases, the first from 1630 to 1637, halted by the English Civil War, with the second phase from 1650 to 1652 completing works that included shortening the course of the River Ouse with major bypass channels. The physical works were undertaken variously by

Scots prisoners but also a virtual army of navvies from Europe's Low Countries, who also brought their names to places across Fenland, from South Holland to Papworth Everard (Everard is a Huguenot name) and many more besides. Today, true fens – low and marshy or frequently flooded area of land – are something of a rarity in the extensive region known as The Fens. By the mid-19th century, the original fens were almost completely destroyed, with the exception of a few small pockets such as Wicken Fen north of Cambridge. (Wicken Fen Nature Reserve, a biological Site of Special Scientific Interest and also a National Nature Reserve, is a wildlife-rich site, yet is in reality a tiny and fragmented remnant spanning 254.5 hectares – little more than 1000th – of the once extensive fens of the Fenlands.) The formerly extensive mosaic of interconnected still and sluggish wetlands ideal for burbot and other wildlife formerly thriving in lowland eastern England was no more, mass drainage securing

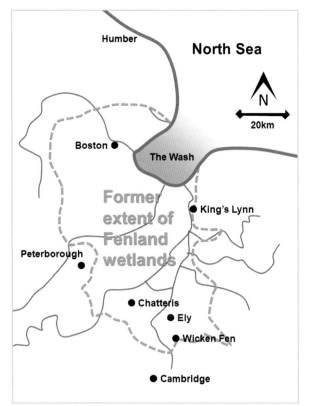

Figure 2.4
Former extent of Fenland in eastern England, true fens now largely reduced to fragmented pockets around Wicken Fen (© Dr Mark Everard)

the terminal decline of the fish here as elsewhere in eastern England and much of lowland Europe.

In *The Lost Fens: England's Greatest Ecological Disaster*, Rotherham (2013) comments that,

> The loss of the great fenlands of eastern England is the greatest single removal of ecology in our history. So thorough was the process that most visitors to the regions, or even people living there, have little idea of what has gone . . . The fenscapes, neither water nor land but something in-between, breed independence and, if necessary, dissention. This story is of politically and economically driven ecological catastrophe and loss. So much has gone, but we do not even know fully what was there before.

Draining of vast swathes for former wetland in eastern England, and with it the eradication of a huge diversity of habitat, biodiversity, livelihoods and traditions, communities and rights, may have been a massively underreported and underappreciated aspect of British history. However, it mirrors much of the even less well documented drainage and habitat loss elsewhere in England and much of lowland Europe, with associated enclosure of former common lands by private interests and consequent eradication of wildlife. The effects of changes further to the north in England are described by Rotherham (2010) in *Yorkshire's Forgotten Fenlands*. In this text, Rotherham discusses the mass drainage and fundamental changes to a wide cultural wetland landscape spanning the Humber basin and the entire county of Yorkshire from the Humber and north Lincolnshire through the Vale of York, through South Yorkshire and Holderness, to Pickering and beyond, once also a stronghold for burbot. Recent decades and centuries have seen profound changes to the wetland landscapes of Britain and Europe, and with it the annexation of human rights and traditions as well as major degradation of biodiversity including extirpation of wetland-dependent species such as the burbot.

Allied with these compound pressures has, of course, been a warming climate, hardly favourable for a cold-water fish. With backwaters and slacker margins vital for the tiny free-floating eggs, dormant for months before hatching out helpless larvae during the cooler months when spate flows are also most likely, and with the larvae living a planktonic life for up to three months, major changes to riverscapes through wholesale landscape conversion can only have had a severely detrimental impact.

Some populations of burbot across their broad northern hemisphere range prosper in deep and cool alpine and prealpine lakes. However, by a strange accident of the geological and climatic past of the British Isles, there are no such lakes connected with the eastern English catchments in which burbot were once present through the former Doggerland land bridge connection. The partial covering to the north and west of Britain by glacial ice means that burbot were never naturally able to colonise the otherwise ideal habitats of the glacial lakes that were left behind as the ice retreated from North Wales to Cumbria and up into Scotland. These naturally nutrient-poor lakes are already under enough pressure from various introduced invasive fish and other species to make the idea of burbot introductions, despite ostensibly ideal habitat suitability, far from a wise strategy.

The former distribution of burbot in eastern England

No formal surveys had been undertaken of burbot before their decline and apparent extirpation. However, there are many references to the former eastern river distribution of burbot in a range of historic texts. For example, Couch (1864) wrote of the burbot in *A History of the Fishes of the British Isles*,

> It is found in the rivers of Yorkshire and Durham, Norfolk, Lincolnshire and Cambridgeshire.

More recently, in a review of burbot captures from the early 19th century up to 1970 where burbot were gathered and arranged watershed-by-watershed in chronological order, Marlborough (1970) concluded that,

Most are from eastward-flowing river systems from Durham southwards to the Great Ouse, but a few records from westward-flowing systems are considered. In many areas the records imply a decline of burbot numbers and distribution during the present century, though burbot may never have been more than locally abundant.

Reviews led by Worthington et al. (2010a and 2011) drew upon evidence from historical literature to map the former distribution of the burbot in British rivers prior to their extirpation, most likely in the early 1970s. These reviews drew on a broad literature comprising around 200 sources, ranging from 12th century farmland practices to 16th century fishing practices, classic texts, as well as a search of scientific papers, and also non-scientific, historical and anecdotal texts. From this collective resource, a scoring system was developed to model anecdotal descriptions of burbot abundance. Forty two rivers in eastern England were identified in which burbot were likely to have existed, ranging from the River Skerne in County Durham to the River Blyth in Suffolk. This eastern English distribution was split into four broad geographical areas, generally related to river catchment boundaries: the Trent catchment and the River Ancholme; Fenland rivers (those draining into The Wash); eastern flowing rivers of Norfolk and Suffolk; and Yorkshire rivers (including the River Skerne in County Durham).

From their reviews, Worthington et al. (2010a and 2011) suggested that burbot persisted in greater numbers in a few isolated areas, such as the Yorkshire Derwent and the Great Ouse. They also concluded that, rather than being confined to slow-flowing lowland rivers, burbot occupied a range of aquatic environments including small upland headwaters as well as some lake systems. This is consistent with the diverse habitat still occupied by the Western European genetic strain of burbot in smaller rivers of continental Europe. As predators, burbot can still prosper in low-nutrient waters, though population size may be limited by food availability. The baseline strength of the British population before this decline was found to be uncertain, its assessment limited by the paucity of reliable documentation relating to the abundance of this fish.

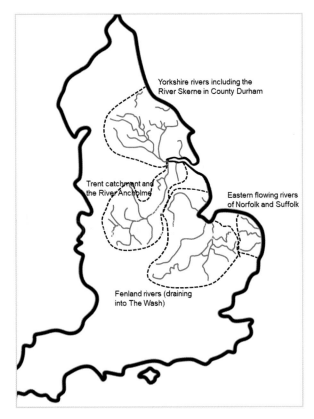

Yorkshire rivers including the
River Skerne in County Durham

Trent catchment and
the River Ancholme

Eastern flowing rivers
of Norfolk and Suffolk

Fenland rivers (draining
into The Wash)

Figure 2.5
Former natural
range of burbot in
England, adapted
from research by
Worthington et al.
(2010a and 2011)
(© Dr Mark Everard)

Evidence concerning burbot in the Thames catchment is unclear, as written sources are contradicted by archaeological records. Four sites across the Thames catchment were found to contain the bones of burbot, though whether this is evidence of the fish being established in the river is uncertain. However, there is no direct evidence of burbot having occurred within the Thames, despite it being connected to the Channel River system during the ice age. Frank Buckland (1881) wrote in *The Natural History of British Fishes,*

> We Londoners very seldom see or hear of a burbolt [sic], and they are such a stupid and ugly fish that I cannot advise trouble to be taken with their dissemination, though doubtless they would thrive in many of our ponds and lakes.

Burbot were not found to have existed in the Thames system according to a report published by Marston (1914). Historical records indicate that the decline of burbot in the Trent catchment, where they were reported as common in the 19th century, occurred from the beginning of the 20th century. This decline was earlier than that observed in the Yorkshire Rivers and the Fenland Rivers. The Fenland Rivers were a stronghold, with most historic records indicating that the Great Ouse catchment supported substantial populations until a decline in the 1940s, and with single captures reported through to the 1960s. The Yorkshire Rivers, including 16 rivers draining to the Humber as well as the River Skerne (a tributary of the Tees) and the Esk (draining directly to sea at Whitby), has scattered records suggesting the species was not uncommon until the early 20th century with the last record from the River Swale in 1967.

Based on their review of extensive literature, Worthington et al. (2010a) summarised a timeline of the prevalence of burbot in their native catchments. This is reproduced in Table 2.1.

Table 2.1 Timeline of the prevalence of burbot in native English catchments (adapted from literature analyses by Worthington et al., 2010a)

Period	Trent	Fens	Yorkshire
Pre-1750	Rare		Often found
1750-	More plentiful		
1800-	Most common	Abundant	Not uncommon
1850-	On the increase	Common	
1870-	Common		Common
1890-		Fairly common	Not so common
1900-	Scarce	Common	Common
1910-	Seldom caught	Many caught	Not rare
1920-		Plentiful	Many taken
1930-		Common	Reasonably common
1950-	Single specimen		Rare
1960-	Single specimen	Unusual catch	A few records

It is interesting that Walton (1653) in *The Complete Angler*, makes no mention of burbot. Since much detail about fish and fishing in that classic work were derived both from Walton's direct experience but also by knowledge gleaned from other anglers, this may suggest that burbot were already scarce or localised in distribution by 1653. In the book *The Natural History of Staffordshire*, editor Robert Plot (1686) wrote that there were only scarce reports of burbot in that county.

We do know that British burbot had dwindled to extinction by the late 1960s. The last recorded capture of a British burbot was a specimen weighing 1 lb, taken at night by John Dean on 14th September 1969 from the Old West River (Great Ouse), near Aldreth in Cambridgeshire. In the 1970s, the weekly magazine *Angling Times* posted a reward of £100 for an authenticated capture of a burbot from UK waters. This prize has never been claimed.

Lingering claims about burbot in Britain

Despite various claims that burbot have been caught since 1969, and the bands of optimists who believe that the burbot may still exist in Britain, there is no confirmatory evidence. Were any still to exist, the most likely places to find remnant populations would be in the larger rivers in the eastern English counties of Yorkshire, Cambridgeshire, Norfolk and possibly elsewhere in East Anglia. The River Cam, a tributary of the Great Ouse system, has been one of the hotspots of unsubstantiated claims.

An article dated 9th July 2010 in the *Angling Times*, a British weekly angling magazine, headlined *'Extinct' burbot spotted in River Eden and Great Ouse,* referenced two reported sightings of burbot swimming in these two rivers, weeks apart at different ends of the country (Angling Times, 2010). Reportedly, these sightings sparked an impromptu netting operation by the Environment Agency; of course, no burbot were found. The reported Great Ouse sighting of a 2 ft long fish, at Needingworth in Cambridgeshire, was by a French angler with a master's degree in fisheries who was familiar with catching burbot in France. The reported sighting on the River

Eden in Cumbria was of a 2 ft 6 in long fish, spotted around the angler's feet for five minutes while he was wading, described as,

'. . . blotchy, with a rounded head, a long dorsal fin and a stubby little tail'.

The western-flowing River Eden is, of course, at a substantial distance from any former catchment with evidence of having hosted burbot. The *Angling Times* also reported in October 2011 that an eight-year-old Yorkshire schoolboy claimed to have caught a burbot from the River Esk near Whitby, though the fish was thrown back so, once again, there is no evidence.

I am minded here of a phone call I received around New Year during this same time period from a friend who fishes the Hampshire Avon, also not a former known home river of the burbot. Breathlessly, he described seeing a longish, largely immobile fish in the margins with a dark appearance; given the post-spawning period, I was thinking 'kelt' (spawned out salmon). He carried on, describing its mottled appearance and how it eventually drifted listlessly out into the main channel, vanishing into the dark water; I was still thinking 'kelt' with a covering of fungal growth as is common at this time of year post-spawning. Finally, he burst out with his identification – 'a burbot!' – verified after he'd gone home and looked at pictures in a book. Not wishing to denigrate my friend, my own analysis is that this was not a burbot! I am sure that the Eden sighting was similar, even though it was in the middle of summer. The reported sighting of the Great Ouse fish is a potentially greater mystery, though unlikely given the absence of unsubstantiated angling captures nor any sign of burbot in repeated fish surveys by fishery management bodies since the 1970s.

One reason that these cryptic fish might remain 'below the radar' may be their nocturnal predatory lifestyle, and their tendency to remain torpid in warmer conditions with peak activity only when the water is cold; river angling using fishy baits overnight through cold snaps in the dead of winter is not a popular pastime! Another reason may be that blind optimism, and possible misidentification, tend to override the fact that there are no burbot to be found!

Chapter 3

Angling for burbot

Lake burbot prefer cold, deep water, coming into shallower water in winter prior to spawning. For this reason, lakes and larger rivers are most productive for burbot fishing when they are frozen. At these times, shallower water is often the most productive as burbot move from deeper layers that become depleted in oxygen and food supplies. However, although burbot are both a sport-fish and are noted as good eating, burbot angling is not generally popular.

Ice fishing for burbot

In many of the large and deep lakes in which burbot thrive across their circumpolar range, they remain inaccessible, far out in the deepest waters for most of the year. It is only during the winter spawning period, from mid-January through to the end of February or into March, that it becomes possible both to get out onto the thick ice and also to intercept burbot as they come inshore to spawn.

Ice fishing methods are popular and broadly similar across the wide geographical range of the fish, though are particularly popular in Alaska, Canada and Finland, and also Russia. Indeed, the burbot is one of the most common species in Finland, occurring through-out the country in all types of waters. Burbot are also common by-catches in North America, in both Canada and northern states of the

United States (US), by anglers ice fishing for lake trout, whitefish, northern pike or walleye (American zander).

Burbot are typically most active at night. In the prolonged and very cold conditions leading to thick ice formation, ice fishing anglers require the right form of angling protection – usually some sort of hut or insulated tent erected around a hole drilled in the ice with an auger – as well as suitable fishing tackle. The most productive periods are related to the crepuscular habits of this fish, which is most active in the first two to three hours following sunset and the hours before sunrise, though bites can occur in the middle of the night. It is also possible to catch burbot during the middle of the day in the deepest, and therefore darkest, water.

Target areas for ice fishing include bays, sounds or headlands over shallower water with a firm bed of sand or gravel on which the burbot will spawn. Fishing is favoured adjacent to rocky areas, as well as in areas around underwater cliffs and the mouths of bays. In large Alaskan lakes, these shallower waters may be 25 ft (just over 7½ m) deep or less. However, in the smaller lakes of Finland, ice over water that is only 1–3 m deep is more common.

When the ice is sufficiently solid and thick, anglers select their spot and drill holes through the ice with an augur that may be manual or powered. In Finland, hardy anglers sit out into the darkening night waiting for a pull on the line or else leave the lines in the water and come back in the morning to see if a burbot has gorged itself. Ice fishers across North America often erect a shanty over the ice hole and spend a number of days fishing it. The author's own experience of this form of ice fishing is that it is a good excuse for a band of fellow enthusiasts to park their trucks out on a lake, escaping the pressures of daily life, camping out with a substantial supply of beer!

There is a further advantage to this communal approach. Burbot can be hard to locate under the ice. So, if several fishermen work together, when one finds burbot under the hole they have drilled, the rest of the fishing party can drill their own holes nearby and exploit the bounty.

The preferred time for burbot ice fishing is at dusk as the fish become more active at nightfall. Set lines can also be used when ice fishing, for which no particularly sophisticated tackle is necessary, or else with standard ice fishing rods (or 'poles' in US terminology) short enough to be used inside the shanties. Tip-up rigs, which flip up a visual indicator when a fish runs with the line, are also popular.

This form of fishing is said to require plenty of concentration, some knowledge of the lake, and a fair bit of luck. Burbot are a challenging fish for the diligent angler. A further tip is that burbot appear to move and forage in small schools at this time of year, so unhooking any caught fish and returning the lure or bait rapidly to the water can be important to intercept more fish when they are present.

Burbot baits

Burbot can be fished for with a wide variety of baits. However, as burbot are distinctly predatory with a preference for fish, live or fresh fish baits are favoured. Most types of fish can be used, including for example whitefish, herring, smelt as well as other fishy baits such as squid. The viscera of other sportfish, such as the head, tail, fins or guts, can be used as bait. However, in the North America state of Alaska, the use as bait of the flesh of other sportfish captured by angling is prohibited.

Ice fishing with live baitfish is a favoured method in the run-up to the spawning period, though dead fish baits are also effective. Preferred baits include live or dead minnows, cut herring, chicken livers or any type of meat that smells strongly enough to attract hungry burbot. Chumming with liquidised fish is often deployed to attract fish into the area in which the angler's bait is presented. Some anglers favour the less ethical but no less effective method for attracting burbot of sinking a couple of dead chickens in a pair of discarded tights which, left for a day or two, are alleged to be highly effective at drawing in foraging fish. Burbot become active at night, most prominently at dusk and dawn, and will often remain near these baited areas for several hours.

Cheese, power baits, worms, salmon roe and even chicken liver can also be used successfully to catch burbot, though generally do not work as well as live or fresh fish baits.

Large baits are generally used as burbot have very large mouths. However, smaller baits can be effective, though it is important that they emit a strong scent. Where lines are set for a long time, it is recommended that the bait is changed at least every day as burbot prefer fresh food.

Lure fishing is also popular, with glowing lures and jigs highly recommended. Applying luminescent paint to lures to make them glow at greater depths is claimed by some to boost the chances of a take by a hungry burbot. Glowing lures are most effective when recharged at intervals with an ultraviolet (UV) light source or a strong torch to retain their attraction in the darker depths. Lures that rattle or vibrate are also claimed to be attractive. The chosen lure is worked through the ice hole and down near the lakebed to resemble a prey fish. Favoured jig rigs include the jigging spoon, often tipped with a minnow tail, or else a leadhead jig tipped with a minnow. These ice fishing lures are rattled along the bottom of the lake or river, although this has to be done slowly as burbot are slow swimmers and they don't chase after swiftly moving lures. The angler works the lure close to the lake or riverbed until they feel the solid resistance of a taking burbot, and then they strike and, hopefully, play the fish up to the ice hole.

Ice fishing tackle

Tackle requirements are not exacting as burbot are not fussy feeders. All that is required is a decent-sized, strong-smelling bait and strong tackle to control it. For all types of ice fishing, the line, be it braided or monofilament, is selected at a breaking strain appropriate for the anticipated size of the fish. Strong lines, between 10 lb and 50 lb breaking strain, but more generally between 10 lb and 20 lb, are required to control a potentially heavy fish.

Hand lines may suffice, but short ice fishing rods are more commonplace. Dedicated ice fishing rods are generally less than 700 mm (almost 28 in) long. Many come equipped with integral reels, whilst

Figure 3.1 Ice fishing rods are short, suitable for lowering lines through ice holes, either with integrated reels or seats to accept separate reel (© Dr Mark Everard)

others have seats to accept separate reels. Fixed spool, small multiplier or centrepin reels can all be used.

Another common piece of ice fishing tackle is the tip-up. A tip-up is a device that signals visually when a fish grabs onto the line. As the name suggests, a tip-up device straddles the drilled ice hole on crossbeams and is equipped with a flag that springs, or tips, up as a fish takes the bait pulling on the line spooled on the device. A variety of types of end tackle can be used depending on the bait type.

Multiple tip-ups are positioned over a number of different holes bored through the ice, generally surveyed by the angler sheltering from the elements in some form of portable pop-up shanty. The angler then waits for one of the tip-up's flag to fly up, the spool spinning freely presenting no resistance betraying the rig to the fish. The angler then grasps the line and pulls sharply to set the hook, playing the fish in the water before landing it by pulling it up through the ice hole. It is sometimes recommended that anglers carry a larger auger to bore out a bigger hole if necessary, as a very large burbot may get stuck in a standard fishing hole.

Figure 3.2 A tip-up rig set over an ice hole cut with an auger
(© Brad Dokken)

In his essay *Ice Fishing, the Moronic Sport* the US writer Jim Harrison (1988) famously stated that,

> The true force behind ice fishing is that it is better than no fishing at all.

Despite this, proponents of the technique argue that adapting tip-up fishing to different species and conditions requires some finesse.

Competitive ice fish

One of the places where competitive burbot fishing occurs is in Flaming Gorge Reservoir, an impoundment retained by the Flaming Gorge Dam on the Green River in the US. Most of the Flaming Gorge Reservoir is in the south west of the state of Wyoming, constituting the largest reservoir in the state, though part of it is in north-east Utah. It is here that the Flaming Gorge Chamber of Commerce promotes an annual 'Burbot Bash', where anglers complete to 'Catch the Ugliest Fish in the West'. Invitations have included the effusive strapline, 'Public is welcome! No entry fees!' The stated purpose of the event was 'To teach beginning ice anglers why and how to catch Burbot also known as ling cod on Flaming Gorge and to put as many of these invasive predators on the ice as possible!'. The event was not, however, all about fishing, with linked events including drawing the fish, as well as communicating tips for gutting, cleaning and filleting fish and other game. Perhaps setting the cultural tone of the event as well as the reality of ice fishing, the concluding prize was the 'Odiferous Award' to be won by the ice fisher that always gets 'skunked' (a term that can relate as much to persistently catching nothing as it can to getting drunk!). The homepage for the burbotbash.com website contains guidance on ice fishing for burbot which is reproduced in Box 3.1.

Every February the town of Walker, Minnesota, holds an annual International Eelpout Festival on the unpromisingly named Leech Lake. Leech Lake is known for its exceptional walleye fishing, the eelpout festival promoting fishing for the often-neglected burbot that inhabits the lakebed. The first International Eelpout Festival was held in February 1979, and it ran annually through to 2019. It took place during one of the coldest periods of the year, when temperatures dropped well below freezing point. The International Eelpout Festival catered for a wide range of audiences and included family activities. The festival received national attention on 4th March 2011 when a correspondent from the television programme, *The Tonight Show with Jay Leno*, recorded a segment on the event. However, the festival was scrapped in 2020 because of pollution

> **Box 3.1   Ice fishing tips for catching Burbot (adapted from the 2021 burbotbash.com website)**
>
> ➤ Auger 20–30 or more holes before wetting a line.
> ➤ If you fish a hole for more than 15 min without a bite – try a different hole.
> ➤ Remember on Flaming Gorge, you can fish up to six rods or tip-ups per angler through the ice.
> ➤ Fish within a few inches of the bottom in 20 to 60 ft, these depths tend to be productive.
> ➤ Jig at least one lure lightly, bouncing the lure on the bottom from time to time.
> ➤ Check tip-ups every 20 min or so, Burbot frequently swallow lures without triggering the flag.
> ➤ Hand jig tip-up lines for a few minutes every time you check them.
> ➤ Don't fish by other groups of people – noise and commotion can negatively impact the bite.

concerns. The festival board is exploring the potential relocation of the event to Lake Bemidji, an hour north west of Walker.

Set line fishing for burbot

In some areas of the North America state of Alaska, set lines are allowed. This trapping technique entails the angler setting bait on a fixed line and leaving it to fish overnight, the fish hooking themselves. Where permitted, set lines can be effective all year round, though are most often set through holes drilled through thick lake ice during the winter months.

Set line fishing, though, is subject to Alaskan state regulations, which vary from water body to water body. A sport fishing license to fish for burbot is required, and each set of kit has to be labelled with the licence-holder's name and address. The set must be checked at least once every day, the hook must rest on the bed of the water body, often with the aid of a heavy lead, and large hooks must be used to avoid snaring and killing undersized burbot and non-target species.

In continental Europe, burbot are reportedly also often caught on eel-lines set overnight.

Summer fishing for burbot

As the fish retreat to deeper water and the ice thins and melts away for the rest of the year, similar rigs and baits are used by a few intrepid anglers in the summer. However, these rigs are deployed in far deeper water and without the benefit of thick ice on which to sit. Drifting live or dead minnows in water deeper than 50 ft (15 m) across the bottom of a lake or deeper river from a boat is often also productive in these warmer times of the year.

However, for most North American and Scandinavian burbot anglers, ice fishing is often the only method used. Also, at least in Finland, burbot are shunned as a game species in summer in preference for other target species, even though their flesh is said to be at its best at that time.

The (non-)fighting burbot

Owing to their predominantly nocturnal habits and slow movements, burbot are not considered very exciting sport as they fight only poorly. Indeed, they are commonly thought to fight the least pound-for-pound of any fish. However, they may grow large and can put up solid resistance, necessitating stronger tackle.

This slow and generally poor fight may explain why burbot are held in low regard by many anglers. This perception is hardly boosted by the unappealing look of the fish, and also their tendency to roll their eel-like tails around the arm of the angler while squirming when the captor attempts to unhook the fish.

Specimen burbot angling

A big burbot, as long as 5 ft (1.5 m), is a hefty beast. Burbot are noted for their solid resistance, rather than a fast or agile fight. Sturdy tackle is required not only as a large and irritable codfish is quite a hefty lump but also to present a large fish, worm or other bait.

If not ice fishing, with its specific techniques and equipment, heavy carp or mahseer tackle comprising multiplier reels or large fixed spool reels with strong lines and hooks may be appropriate. The same gear may be used for presenting static baits or using jigs.

Burbot as bait

Larger burbot are predators of smaller fish. But equally, larger predatory fishes will readily take smaller burbot. Predators range from opportunists feeding on abundant pelagic burbot larvae, through to bigger predatory fishes eating progressively larger burbot. This may include a considerable degree of cannibalism where burbot of significantly different sizes encounter each other.

Whilst burbot seem not to be specifically targeted as a live or dead bait by anglers pursuing predatory fishes, a number of manufacturers market artificial lures imitating burbot.

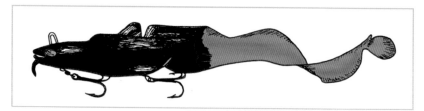

Figure 3.3 Lures imitating burbot are available for predator anglers (© Dr Mark Everard)

Pursuing English burbot

There are two significant obstacles that the English burbot angler must overcome. Firstly, there is no thick winter ice upon which to walk and through which to cut a hole to present baits or jigs. Secondly, and perhaps more significantly, there are no burbot! However, far be it from the intrepid English burboteer to be put off by such trifles!

A significant difference to a potential British approach to burbot angling is that we do not have the very large lakes nor the long-term low temperatures that create thick ice cover for which ice fishing techniques have been developed. Rather, most of our former burbot appear to have been fish of rivers and connected streams and wetlands that were not prone to freezing over with any regularity or long duration. These conditions also prevail across much of the southernmost European distribution of the burbot, for example in northern Italy as well as in northern France and Germany. Traditional ice fishing rods can be discarded, and anyhow may be illegal in many parts of the British Isles where minimum rod lengths often apply as a measure to control poaching.

Burbot were caught in eastern England within living memory. Testimony directly from some now much older anglers, as well as further handed-down verbal and written observations, suggests that, at least until the 1950s and early 1960s, the capture of a burbot was not greatly remarked upon. Neither were techniques specialised, burbot turning up in mixed catches as anglers deployed basic float and leger techniques. Admittedly, sophisticated specimen angling was in its infancy in this era, but no specific tactics were talked and written about with respect to the burbot.

Based on what we know now about both the fish and evolving angling methods, long-stay legering tactics, particularly in winter and with fishy or meaty baits, would appear to be the most appropriate approach to intercept any lingering burbot by design as they migrate into shallower water to feed as night falls. Equally, putting a smelly fish bait near crevices or woody debris adjacent to the bank of a river in which a burbot may lurk would seem a sensible approach. In practice, heavy tackle backed up by electronic indicators akin to that used when pike, catfish, carp or eel fishing, presenting large and smelly fish-based baits, may be effective for any British burbot that has forgotten that it is extinct. Large fixed-spool reels armed with line of 10–20 lb breaking strain, attached to pike or carp rods with a test curve of 1.75–2.5 lb, should fit the bill for these fish with a reputation as poor fighters.

Large baits seem to be preferred, including for example bunches of minnows or chunks of other larger fish, chicken livers or strongly smelling meat of any other kind. Other less orthodox baits may be appropriate in a British context, particularly those that are fish-based and preserved to increase their smell and to help them stay on the hook for the long stays required when waiting for a biting burbot.

Hook type need not be a major consideration, though a large hook, for example from size two to four depending upon the size of the quarry, is required for large baits. Interestingly, as late as the 1950s, Norwegians still used juniper wood to craft burbot hooks. This may seem anachronistic as we are familiar today with ubiquitous cheap, strong and sharp metal hooks. However, fishing hooks used by human societies around the world over thousands of years were crafted from a range of materials including wood, animal and human bone, horn, shells, stone, bronze and iron. Quality steel hooks began to make their appearance in Europe in the 1600s, and hook-making became a task assigned to professionals.

Legering tactics would seem to be most appropriate as evidence from other regions in which burbot are still fished for suggests that a bait firmly on the bed of the river or pool is necessary. Legering tactics are also amenable to static 'bait-and-wait' tactics when paired with electronic indicators. The leger should run freely on the line, and the rod should be set on rests with the reel's bail arm open to allow the fish to run unimpeded, offering absolutely minimal resistance to the running fish as for example when eel fishing. Drop-off bite alarms or other electronic indicators to wake or alert the (very) long-stay angler are the order of the day. When (or if) a run occurs, the angler should close the bail arm and wind down to set the hook, avoiding deep hooking. Given that pike are likely to pick up these baits if they are present, soft wire or Kevlar traces are the order of the day if the line is not to be bitten through leaving the hooks in the unfortunate pike. It seems that the burbot is an unfussy feeder, and hence unlikely to be put off by a bit of wire.

Given the burbot's habit of lurking in cover, the bait should be cast into the deepest part of large still waters, as well as near cover, or else near deep tree roots, undercut banks or other similar haunts in large rivers thought (or at least imagined) to hold burbot.

Since 1969, the point at which the last authenticated burbot was caught from the British Isles, there have been many angling expeditions to capture and authenticate a British burbot. Also, with pike angling becoming popular particularly since the 1980s, any remnant winter-feeding British burbot would have been likely to have encountered fishy baits thus presented at potential ambush points.

Notwithstanding various rumours and unsubstantiated claims, it is generally accepted by most people that the burbot is extinct within British shores. Therefore, the manner in which one might go about fishing for them should pay a greater attention to style than stealth. For example, a large float is more stylish than long-stay carp tackle, as is wearing a suitable hat not only to keep the sunlight or moonlight out of your eyes but also to cut a dash on the river or lake shore!

For me, I think I would prefer to trot or quivertip bread in a winter river for burbot, enjoying the accidental by-catch of chunky roach and dace, or else to present sweetcorn by lily pads under a half-cocked float in summer and enjoy sport with tench or rudd whilst I wait for Godot, or for burbot, to turn up, whichever does so soonest!

For the United Kingdom (UK) resident seriously intent on catching a burbot, the best option is to book a guided trip with those that cater for such things in Alaska, Canada, Finland, Denmark or any other burbot fishing hotspot!

World record rod-caught burbot

The International Game Fish Association (IGFA), generally considered the world's governing body for sport fishing, keeps a current list of World Record rod-caught fishes. At the time of writing

(February 2021), the IGFA World Record burbot is a fish of 25 lb 2 oz (11.4 kg) caught on 27th March 2010 from Lake Diefenbaker in Saskatchewan, Canada, by recreational angler Sean Konrad.

However, the North America state of Alaska recognises a burbot record of 24 lb 12 oz (11.2 kg), taken in 1976 from Lake Louise by a George Howard.

Another claim for the world record burbot is a fish weighing 22 lbs 8 oz (10.2 kg) caught by Vaughan Kshywiecki from Lake Athapapuskow, Manitoba. This immense fish, measuring 1.08 m (42.5 in) long, was taken during the Flin Flon Fish Enhancement Society's Burbot Derby.

British record rod-caught burbot

British record rod-caught fish are recorded and verified by the British Record (Rod-Caught) Fish Committee. However, there is no official British rod-caught record for the burbot. In fact, no claims for records for the species will be entertained by the Committee due to the rare or threatened species status of the fish under the provisions of the Wildlife and Countryside Act (1981) and later Orders. Under this legislation, burbot join allis shad (*Alosa alosa*), schelly, powan or gwyniad (*Coregonus lavaretus*), common sturgeon (*Acipenser sturio*) and vendace (*Coregonus albula*). However, historic, pre-legislative British records remain in place for allis shad and schelly.

The last verified burbot of 1 lb, caught on 14th September 1969 from the Old West River (Great Ouse) near Aldreth in Cambridgeshire, is therefore something of a default unofficial British rod-caught record.

However, many larger burbot are reported as having been caught historically from British rivers. Regan (1911) wrote in *The Freshwater Fishes of the British Isles* that the largest recorded specimen in the British Isles was an 8 lb burbot from the River Trent saying,

> As a rule in our waters it does not attain a length of more than 2 feet, with a weight of about 3 lbs; a specimen of about 8 lbs from the Trent seems to be the largest recorded English Burbot, but double

this size is recorded on the Continent, and in the arctic regions they grow very large, Burbot weighing as much as 60 lbs having been taken in Alaska.

Potentially referring to the same British fish, the Reverend W. Houghton (1879) in *British Fresh-water Fishes* gives more details,

According to Pennant, the largest British specimen ever heard of was taken in the Trent by Sir Gervase Clifton, which weighted eight pounds, a fish of very unusual size.

A final cautionary note to the burbot angler

The tendency of burbot to roll their eel-like tails around the arm of an angler attempting to unhook them has already been noted.

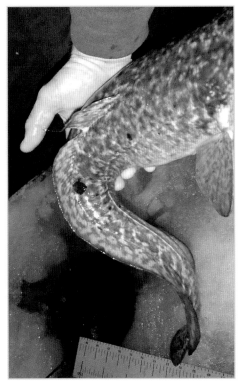

Figure 3.4 Captured burbot, like this specimen captured in the dark, tend to roll their eel-like tails around an angler's arm (© Damien Collins)

To this, A. Laurence Wells (1941) adds a further cautionary note in *The Observer's Book of Freshwater Fishes of the British Isles*,

> In the adult the mouth is well supplied with rows of sharply pointed teeth which it has no hesitation in using to good purpose. The angler who catches them occasionally, when bottom fishing with worms, may discover this to his cost.

Chapter 4

Cultural connections with burbot

The burbot is far more than simply a target for anglers. There are also many diverse cultural connections to this fish, some of which will be explored in this chapter.

What's in a name?

The word 'burbot' is derived from the Middle French *bourbotte*, meaning to wallow in mud. However, burbot are also known by a wide range of other names across their broad geographical range.

Alternative English names include the burbolt, eelpout (not to be confused with the marine eelpouts of the family *Zoarcidae*), and ling (again not to be confused with any of several marine members of the hake family *Lotidae* within the genus *Molva*). In the east of England, the names 'poult' and 'powte' are used. The name 'freshwater cod' is also applied to this fish, along with bubbot, lingcod, freshwater ling, loache, methyl, lush, gudgeon (many other types of fish are also known by this name), mud-blower, cusk, mother eel, lota and mariah. 'Lawyer fish' is another name used for the burbot, under some definitions owing to the fish's 'beard', though other less charitable observers state that this is linked to the slimy skin and slippery nature of the fish!

Mammalian analogies also feature in local names, as described by Houghton (1879) in *British Fresh-water Fishes*,

Figure 4.1 Wild adult burbot on gravel in Khuvsgul Lake, Mongolia (© Batbold Tsedensodnom)

. . . the habits of this fish are to conceal itself under stones and deep banks, and on this account it has been called Coney or Rabbit Fish from its lurking nature.

Houghton (1879) also notes that Pliny the Elder (Gaius Plinius Secundus (23–79 AD) the author, naturalist, natural philosopher and military commander of the early Roman Empire) named the fish after the weasel, perhaps also after its 'lurking' nature, in his *Naturalis Historia*.

This curious fish appears to have been known to, or at least mentioned by Pliny under the name of mustela, or 'weasel fish'.

In the 1686 book *The Natural History of Staffordshire* edited by Robert Plot, it is noted that,

. . . by others, from the oddness of the shape, and rarity of meeting them, the Non-such; there having never but four (that I could hear of) been found within memory.

In the United States (US), burbot are sometimes known as the American burbot. Referring to the Kootenai river system rising in the Canadian state of British Columbia, and running southwards through the US state of Idaho, the Idaho Department of Fish and Game (IDFG) refers to the burbot as 'The Leopards of the Kootenai'.

As we have seen, the genus and species name *lota* comes from the French *lotte de rivière* (the old French word *lotte* meaning 'monkfish').

Other names by which burbot are commonly known in the languages of the countries in which the fish is found are shown in Box 4.1.

Box 4.1   Showing the names by which the burbot is known in the native language of other countries.

Language	Local name
Swedish	Lake
Finnish	Made
Russian	Nalim (налим)
Mongolian	Gutaar (Гутаар in Cyrillic)
Estonian	Luts
Lithuanian	Vėgėlė, Burbotas
Latvian	Vēdzele
Polish	Miętus
German	Quappe, Trüsche, Aalrutte
Danish	Knude, Ferskvandskvabbe, Løgl
French	La Lotte, Lotte de rivière
Dutch	Kwabaal

Although the accepted scientific Latin name of the burbot is now *Lota lota* (including in two subspecies names as we have noted previously), this has superseded a number of prior scientific names. Historically, burbot have been known by other Latin names including: *Gadus lota*, *Gadus lacustris*, *Gadus maculosus*, *Gadus compressus*,

Lota vulgaris and *Lota marmorata*, amongst many others, including some with additional subspecies names.

A further interesting connection with one of the burbot's local names is that of the village of Whaplode, a village and civil parish in the South Holland district of the English county of Lincolnshire. In *Our place: can we save Britain's wildlife before it is too late?* Cocker (2018) writes,

> Another intriguingly fish place that spoke of wild abundance was the village just west of Gedney Fen called Whaplode. This odd name, appropriately enough, comes from that strange, ugly, snake-skinned, freshwater relative of the cod call the burbot *Lota lota*.

Cocker (2018) goes on to clarify,

> In the Fens and elsewhere the burbot was better known by a variety of local names (coney fish, barbolt, ling, eelpout). In the case of Whaplode, however, which means roughly 'burbot stream', the first syllable derives from the earlier alternative name, the quap or quab, whose root was a verb meaning 'to quiver, flop, palpitate or throb' (describing presumably the burbot's rather eel-like action), whence also the word 'quagmire'.

Die Quappe is one of the German names of the burbot, possibly entering the local lexicon as a result of the extensive involvement of navvies from the Low Countries in the draining of the Fens either side of the English Civil War. This phase of major engineering under the guidance of Dutch engineer Sir Cornelius Vermuyden was one amongst many incremental initiatives progressively desiccating the former rich wetlands of Fenland from Roman times through to the 1950s in the name of agricultural improvement. Collectively, these incrementally eradicated former local communities, rights and traditions, along with rich aquatic biodiversity, including the previously widespread and commercially exploited burbot.

Burbot gastronomy

Burbot feature in many medieval recipes, including pies and soups, and were reportedly also formerly highly prized by French chefs and the tsars of Russia.

Waterzooi is a stew dish from Belgium, originating in Flanders. It is sometimes called Gentse Waterzooi, referring to the town of Ghent where it was said to have originated. Whilst the '–zooi' element of the name derives from Middle Dutch words relating to boiling, the original dish was made of fish (known as Viszooitje) and in particular burbot. However, it is believed that, as pollution of the rivers around Ghent led to the disappearance of burbot, alternative fish of either freshwater or marine origin replaced them. Today, eels, pike, carp and bass are commonly used, though other fish such as cod, monkfish or halibut can also be used. However, the most common form of waterzooi today contains chicken (known as kippenwaterzooi), replacing the fish. It is said that Charles V, the Holy Roman Emperor in the 16th century, was keen on eating this rich fishy dish despite suffering from gout (Carson, 1978).

Further back in history, Pliny the Elder (23–79 AD) in his *Naturalis Historia*, written during the time of the early Roman Empire, was complimentary about burbot. Pliny is quoted by Houghton (1879) in his book *British Fresh-water Fishes* as relating that,

> The next fish (to the *scarus*) best for the table are the *mustela*, which, strange to say, the lake of Brigantia, in Rhætia, amongst the Alps, produces, rivalling the fish of the sea.

Burbot are reputed to have excellent white, flaky flesh, said to be tasty if a little dry. Some people find this odd given the unattractive appearance and slimy feel of the live fish. Sometimes, it is recommended that, as for other members of the order of codfish, burbot are gutted and bled shortly after capture to preserve the quality of the flesh.

The fish is also best skinned. Removing the skin with a knife is a difficult process. The Sport Fish Division of the Alaska Department of Fish and Game (ADF&G), offers an illustrated online guide to the skinning and filleting of burbot (see bibliography for website address). This guide includes a series of steps covering just one method of skinning that commences with cutting the skin behind the head, then grasping it with pliers and pulling towards the tail. With the skinned burbot on a slab, the rear of the dorsal fin is grasped with the pliers and pulled towards the head, removing the dorsal fin from the body. With the fish then flipped onto its back, this process is repeated to remove the anal fin. With the fish turned back onto its belly, a knife is inserted where the dorsal fin has been removed and is run down along the ribs to remove fillets from both flanks. The belly meat is prized and can be extracted after the flank fillets have been removed.

There are many recipes for burbot, particularly from Alaska and Canada where the fish is widespread and popular as a sport fish. Many of these recipes can be applied generically to the white flesh of any of the families of codfish. Cooking methods include many variants of boiling, poaching, baking, deep-frying, pan-frying and even pickling. Burbot flesh is also often smoked. Indeed, it is also said to be the best of all smoked fish species.

When sold, the flesh of the burbot is mainly salted. However, some is canned, after which it is rumoured to taste like crabmeat. Indeed, some describe cooked burbot meat as similar to American lobster, leading to a popular nickname for burbot as the 'poor man's lobster'. A good many recipes for burbot use this fish as a substitute for the exorbitantly priced lobster. This includes, for example, 'Burbot Newburg' (a variant of the classic 'Lobster Newburg' dish) which entails adding two cups of raw burbot chunks to melted butter (4 tablespoons) and sherry (a quarter cup) until the alcohol has evaporated, after which a quarter-teaspoon of salt is added. In a separate bowl, two beaten eggs are blended into one cup of cream, and this is then stirred into the fish mixture. A dash of cayenne is added and the whole mixture is then cooked and stirred slowly over moderate heat until the sauce thickens. *Bon appétit!*

Whilst burbot flesh has not found universal acceptance, burbot livers are widely considered a delicacy. Indeed, Montagne (1938) in *Larousse Gastronomique* states that,

A woman would sell her soul for a burbot's liver . . .

Figure 4.2 Burbot flesh and livers (© Joachim Claeyé/AquaLota)

Buckland (1881) mentions in *The Natural History of British Fishes* that,

The flesh is said to be good, especially the liver when fried, but it is indigestible.

The liver of a burbot can account for some 8–10% of the overall body weight of a wild fish, or as much as 15% in aquaculture. This proportion in wild burbot is up to six times the size of the liver of many other freshwater fish of comparable size. Burbot livers can be found for sale smoked or canned in Europe.

In Finland, the roe of the burbot is sold as caviar. Indeed, Finnish burbot caviar is well-renowned. Some gourmets claim that it is the highest quality caviar of them all, and is finer than sturgeon in both taste and cost. However, the roe of burbot, as for pike, perch, ruffe and other popular Finnish sources of caviar, has to be frozen before consumption to prevent the passing on of tapeworms to the consumer.

Burbot are enjoyed for food in the North American state of Alaska, where the Sport Fish Division of ADF&G, has also published a

Figure 4.3 Burbot recipe book published through open access by the Government of Alaska's Department of Fish and Game, Sport Fish Division

book titled *Burbot Recipes*. This is available freely online (see the bibliography for the website address), so I will not reiterate all of its fine details. Suffice to say that it is a 'go to' work for anyone keen to deepen their knowledge of 'top-of-the-stove' and oven recipes for burbot. The aim of this Alaskan guide is to familiarise people with this little-known fish, admittedly unattractive in appearance, but which has a large, vitamin-rich liver and a firm, flaky flesh not dissimilar to that of other codfish species taken from ocean waters.

Religion and the burbot

An interesting aspect of the embedded nature of the scales of the burbot, and their smooth and apparently scaleless appearance, is that it creates particular difficulties for some religious groups.

For the Jewish faith, in order for a fish to be kosher – food that complies with the strict dietary standards of traditional Jewish law – it must have two signs given in the Torah. (The Torah – the 'Instruction', 'Teaching' or 'Law' – is determined to be the first five books of the Hebrew Bible, or Pentateuch, though can also mean the totality of Jewish teaching.) These two signs are that the fish must have fins and scales. To prove that a fish is kosher, the scales must be visible to the naked eye and must be easy to remove from the skin of the fish, either by hand or with an instrument. Although the burbot has scales, these are small and deeply embedded, and so it is considered a 'skin fish' and therefore not kosher.

Similar considerations apply in the Islamic faith in terms of fish that are considered Halaal (allowed) and those that are Haraam (forbidden). Whilst the codfishes are almost exclusively Halaal, burbot (generally referred to as 'freshwater cod') are considered Haraam.

The medicinal burbot

The medicinal values attributed to cod liver oil, a commonplace food supplement available from health food shops, are well known. It should not then be entirely surprising that oil from the liver of the burbot, a fish of the Gadiformes order of cod-like fishes, might be ascribed similar qualities.

In their research paper, Clow and Martlatt (1929) explored the use of burbot liver oil as an antirachitic, or in other words for the treatment of rickets. Rickets is the condition of weak or soft bones in children, typically involving bowed legs, stunted growth, bone pain, large forehead and trouble sleeping, though potentially with wider, serious consequences. The primary cause of rickets is vitamin D deficiency, before and after birth, though there may also be hereditary factors. Today, the treatment for rickets includes increasing the dietary intake of calcium, phosphates and vitamin D. Exposure to ultraviolet-B light promotes vitamin D synthesis, most easily obtained by exposure of the skin to sunlight. Cod liver oil and halibut liver oil are two recognised sources of dietary vitamin D. Burbot livers have more recently been found to contain a high concentration of vitamin D, vitamin K, vitamin A and polyunsaturated fatty acids.

As the liver of the burbot is large compared to that of other fishes, it has the potential to be used as a suitable alternative source of vitamins D and K to other cod-like fishes. This provides a potential commercial use for the burbot livers, with the flesh of the fish being available as a food source.

Commercial burbot fisheries

There seem not to be, nor have ever been, any commercial fisheries dedicated to burbot as a food fish. As Wheeler (1969) describes in *The Fishes of the British Isles and North West Europe*,

> Its flesh is said to be good eating, and the liver is as rich in vitamin A as that of the cod. Despite this, the burbot is very little exploited even where still common.

At one time, Canada did try to launch a commercial burbot fishery, exploiting both the oil from the liver and the meat, but it did not take off. In the main, burbot are a by-catch of traps and nets set for perch, walleye, bass and other more desirable and economically valuable species.

Indeed, burbot are sometimes considered a nuisance to commercial fisheries, eating other commercial fishes caught in gill nets, or clogging gill nets and wasting the time involved in their removal.

However, burbot by-catch from commercial fisheries in the 1920s formed the basis for what was to become a major chemicals company. In 1920, Joseph Rowell returned from service in the First World War to settle on the shores of the Lake of the Woods in northern Minnesota. Initial attempts at fishing for more desirable species were moderately successful, though the fishery's nets were consistently filled with a rough fish, the burbot, which had no commercial value. However, Rowell had also cofounded another venture, the Northern Blue Fox Farm, which commercially bred blue foxes for the lucrative fur industry. Turning inconvenience into opportunity, Rowell started feeding the foxes with the unwanted by-catch of burbot, producing healthy foxes with superior coats, which also tended to have far larger litters than many other farmed foxes. Suspecting a linkage between the burbot and the rich fox pelts, Joe Rowell's son Ted applied his pharmacological skills to demonstrate that burbot liver oil was exceptionally high in vitamins A and D. Following this discovery, Ted Rowell developed a method to extract pure oil from the burbot livers, developing burbot liver oil as a marketable product. In 1935, father and son incorporated their fledgling company as the Burbot Liver Products Company, which proved a great success. By 1940–41, the Burbot Liver Products Company purchased half a million pounds (227,000 kg) of burbot livers annually, sourced from all 30 of the commercial fishermen on the Lake of the Woods as well as other fisherman throughout northern Minnesota. Sales were boosted further still by the Second World War, which effectively cut off supplies of cod liver oil that had been largely sourced from Norway. The company also began to diversify beyond burbot liver oil products such that, by 1949, the name of the company was changed to Rowell Laboratories, Inc. The lineage of this company still survives at the time of writing as part of the Belgian company Solvay.

Burbot farming

In his book *British Fresh-water Fishes*, Houghton (1879) wrote,

> This fish is very deserving of cultivation, and I hope that pisciculturalists will soon turn their attention to the burbot.

Burbot have more recently been identified as a potential new species for commercial aquaculture in the US state of Idaho, with commercial production facilities being set up. The fact that they have not been farmed for utilitarian purposes to date may be related to difficulties created by their inherent aggression, predatory diet and requirement for low temperatures. Whilst these are not insurmountable barriers in themselves, the economics of burbot production may be further undermined by low demand for a perceived 'ugly' fish regardless of its other uses and virtues.

Other hatcheries and growing-on facilities have been established in both the US and northern Europe. Mainly, these are for research and potential burbot reintroduction purposes rather than for strictly commercial aquaculture. However, there are two commercial burbot hatcheries in Europe. The LOTAqua hatchery in Germany (see Everard, Chapter 5, this volume, 2021) and the AquaLota hatchery in Zele, Belgium (see Box 4.2).

Box 4.2   The AquaLota commercial burbot hatchery (Flanders, Belgium)

The AquaLota burbot hatchery in Zele, East Flanders, Belgium, was set up by Joachim Claeyé and his father Guido. This followed Joachim's studies in agriculture biotechnology and aquaculture at Odisee University of Applied Sciences. Odisee University started a programme of applied research on burbot culture in recirculating aquaculture systems (RAS) under an Interreg-project titled 'AquaVlan'. AquaVlan was a cross-border collaboration programme with financial support from the European Fund for Regional Development. Burbot culture has remained a major research topic at Odisee University, and Joachim has been involved both as a student and, later, as a research assistant.

50,000 burbot fingerlings were produced in AquaLota's first operational season in 2020 with the aim of increasing production in subsequent years. These fingerlings are produced primarily to serve niche aquaculture markets for human food production: burbot are said to have a taste somewhat similar to monkfish and turbot, and the liver and caviar are also considered delicacies. Broodstock for the AquaLota venture derived from the Central European strain from the Rhine, the Elbe and the Oder catchments in Germany. Fish from these distinct catchments were selected in preference to the Western European strain as they are larger and grow more quickly, and hence are more favourable for food production.

There have been requests in recent years to sell fingerlings for reintroduction programs. In this case, special care is taken to ensure the genetic integrity of existing burbot populations, stocking only the strain appropriate to the target catchment.

A group of 175 broodstock burbot is kept at the AquaLota hatchery to maintain genetic diversity during captive spawning. Burbot larvae demand live feed, which is expensive and time-consuming in aquaculture. Processes have been developed at AquaLota to feed hatchlings from swim-up with phytoplankton and zooplankton for a period of up to two months before graduating to dry granulated feeds. The cannibalistic tendencies of burbot pose further management challenges. Fingerlings are sold at a weight of 10 g when they are approximately 10 cm long. Over the years, AquaLota has developed a reliable protocol yielding consistent and satisfactory results.

Burbot were observed to have a very high food conversion ratio (FCR) of 0.8–1.0, meaning that between 800 and 1000 g of feed is required to produce 1 kg body weight of burbot. Using similar feed, rainbow trout have a FCR of 1.2–1.4. The substantially more efficient conversion ratio seen in burbot is believed to be due to these sessile fish not wasting energy by constantly swimming, and their slow digestion rate which enables them to maximise nutrient absorption.

Howell and Fletcher (2020) provide further details on the AquaLota hatchery and process in their online article *All about the burbot* (see bibliography for web address).

A rich array of scientific literature has been published on the optimal conditions for burbot breeding and rearing from research at the Belgian government's *Instituut voor Natuur-en Bosonderzoek* (INBO) fish conservation hatchery at Linkebeek. Further knowledge has been generated by observations and practices at LOTAqua and AquaLota. Relevant aspects of this research are incorporated into this book though much of the finer technical detail is beyond the book's scope. This wider body of science covers factors such as the different virtues of temperature modification to stimulate maturation, hormone-promoted versus natural maturation of broodstock, and how temperature, light levels and diet used during raising affects larval and juvenile health and growth rate.

We will return to the use of fish conservation hatcheries for potential burbot reintroduction later in this book (see Everard, Chapter 5, this volume, 2021).

Burbot harvesting and use

In their exploration of the potential of burbot liver oils for the treatment of rickets, Clow and Marlatt (1929) reported that,

Figure 4.4 Eggs at different ages in the LOTAqua hatchery, Germany (© Hendrik Wocher)

The U.S. Commissioner of Fisheries reported in 1928 that in the total catch of fish in Lake Erie 170 tons of burbot were taken. As these fish are seldom used for human food, they are either thrown away or used as fertiliser. The annual catch as indicated by the commissioner is 200 tons a year from the Great Lakes. Some of the nearby states have made efforts to destroy the burbot in the streams and ponds where they feed upon the young fry of trout. In the waters of Lake Winnebago in Wisconsin 25 tons were thus removed in 1925–1926. The average weight of the burbot is about three pounds, but the liver is reported to be at least six times as large as that of any other fish in these waters.

This text is reproduced here as it also indicates the scale of burbot harvesting at that time, albeit mainly as by-catch. It also emphasises the low value attributed to the fish, minimally for food, but by large tonnage treated as waste or fertiliser. However, other uses are made of burbot products elsewhere in the world.

Figure 4.5 Juvenile burbot growing on at the LOTAqua hatchery, Germany
(© Hendrik Wocher)

Figure 4.6 Burbot fingerlings produced at the AquaLota hatchery ready for
sale (© Joachim Claeyé/AquaLota)

Burbot have historically been used for animal feed. By some accounts, burbot were once so common in eastern England that they were fed to pigs. Referring to the extreme scarcity or extinction of the British burbot today, Nick Giles (1994) comments in *Freshwater Fish of the British Isles: A Guide for Anglers and Naturalists* that,

> This is a distinct change from the situation in the sixteenth century when burbot were so abundant in certain East Anglian Fenland rivers that they were scooped out and fed to pigs as a cheap, high-protein food source.

In addition to food, and as a source of vitamins discussed previously in this chapter, commercially captured burbot in Canada, generally as by-catch rather than as a target species, were mainly rendered into fish meal (Scott and Crossman, 1973). This generic fishmeal, with burbot processed along with a range of other less commercially valuable fish species, may be used as food for farmed animals including mink and other species raised for fur.

Rather more intriguingly, it is said that the skin of the burbot is used in Switzerland to produce a fine, soft, beautifully patterned leather for watchbands, wallets and other accessories (Fantom, 2020). Burbot skin is also used in Russia as stated by Dahlstrom and Muus (1971) in the *Collins Guide to The Freshwater Fishes of Britain and Europe*,

> Even the skin is tanned in Siberia, and put to various uses.

In a review of ethnoarchaeological uses of fish skin, Vávra (2020) notes that burbot skin was in fact quite widely used. Many of these uses were similar to those of eel-skins, including as flail-links, bags or purses. More specific uses of burbot skin listed by Vávra include the mending of cracks in glass windowpanes, or as coverings for cracked bottles using burbot skin flayed in one piece. The Khanty people of Siberia were reported to have used burbot skin as windowpanes. Vávra also reports archival photographs of a Cree man from Saskatchewan making a rattle out of fish skin, either burbot or pike skin being used for this purpose.

Burbot as an inspiration of great art

The book *Anna Karenina* by the Russian author Leo Tolstoy (1878) (Count Lev Nikolayevich Tolstoy: 1828–1910) includes a mention of burbot as table food,

> 'Ah, "Alines-Nadines" to be sure! There's no room with us. Go to that table, and make haste and take a seat,' said the Prince, and turning away he carefully took a plate of burbot soup.

Anton Pavlovich Chekhov (1860–1904), the famous Russian short-story writer, playwright and physician, is generally considered one of the greatest short-story writers in world literature. He was also a very keen angler, writing about ruffe, gudgeon, chub, perch, carp as well as burbot. Indeed, one of his comic short stories concerns the exertions of two Russian peasants trying to dislodge a stubborn burbot from submerged tree roots. Since it is both enjoyable and classic literature, an English translation of the tale is reproduced below from the original Russian "Налим" (Chekhov, 1885),

> A SUMMER morning. The air is still; there is no sound but the churring of a grasshopper on the riverbank, and somewhere the timid cooing of a turtle-dove. Feathery clouds stand motionless in the sky, looking like snow scattered about . . . Gerassim, the carpenter, a tall gaunt peasant, with a curly red head and a face overgrown with hair, is floundering about in the water under the green willow branches near an unfinished bathing shed . . . He puffs and pants and, blinking furiously, is trying to get hold of something under the roots of the willows. His face is covered with perspiration. A couple of yards from him, Lubim, the carpenter, a young hunchback with a triangular face and narrow Chinese-looking eyes, is standing up to his neck in water. Both Gerassim and Lubim are in shirts and linen breeches. Both are blue with cold, for they have been more than an hour already in the water.
> 'But why do you keep poking with your hand?' cries the hunchback Lubim, shivering as though in a fever. 'You blockhead! Hold

him, hold him, or else he'll get away, the anathema! Hold him, I tell you!'

'He won't get away . . . Where can he get to? He's under a root,' says Gerassim in a hoarse, hollow bass, which seems to come not from his throat, but from the depths of his stomach. 'He's slippery, the beggar, and there's nothing to catch hold of.'

'Get him by the gills, by the gills!'

'There's no seeing his gills . . . Stay, I've got hold of something . . . I've got him by the lip . . . He's biting, the brute!'

'Don't pull him out by the lip, don't – or you'll let him go! Take him by the gills, take him by the gills . . . You've begun poking with your hand again! You are a senseless man, the Queen of Heaven forgive me! Catch hold!'

'Catch hold!' Gerassim mimics him. 'You're a fine one to give orders . . . You'd better come and catch hold of him yourself, you hunchback devil . . . What are you standing there for?'

'I would catch hold of him if it were possible. But can I stand by the bank, and me as short as I am? It's deep there.'

'It doesn't matter if it is deep . . . You must swim.'

The hunchback waves his arms, swims up to Gerassim, and catches hold of the twigs. At the first attempt to stand up, he goes into the water over his head and begins blowing up bubbles.

'I told you it was deep,' he says, rolling his eyes angrily. 'Am I to sit on your neck or what?'

'Stand on a root . . . there are a lot of roots like a ladder.' The hunchback gropes for a root with his heel, and tightly gripping several twigs, stands on it . . . Having got his balance, and established himself in his new position, he bends down, and trying not to get the water into his mouth, begins fumbling with his right hand among the roots. Getting entangled among the weeds and slipping on the mossy roots he finds his hand in contact with the sharp pincers of a crayfish.

'As though we wanted to see you, you demon!' says Lubim, and he angrily flings the crayfish on the bank.

At last his hand feels Gerassim's arm, and groping its way along it comes to something cold and slimy.

'Here he is!' says Lubim with a grin. 'A fine fellow! Move your fingers, I'll get him directly . . . by the gills. Stop, don't prod me with your elbow . . . I'll have him in a minute, in a minute, only let me get hold of him . . . The beggar has got a long way under the roots, there is nothing to get hold of . . . One can't get to the head . . . one can only feel its belly . . . kill that gnat on my neck – it's stinging! I'll get him by the gills, directly . . . Come to one side and give him a push! Poke him with your finger!'

The hunchback puffs out his cheeks, holds his breath, opens his eyes wide, and apparently has already got his fingers in the gills, but at that moment the twigs to which he is holding on with his left hand break, and losing his balance he plops into the water! Eddies race away from the bank as though frightened, and little bubbles come up from the spot where he has fallen in. The hunchback swims out and, snorting, clutches at the twigs.

'You'll be drowned next, you stupid, and I shall have to answer for you,' wheezes Gerassim. 'Clamber out, the devil take you! I'll get him out myself.'

High words follow . . . The sun is baking hot. The shadows begin to grow shorter and to draw in on themselves, like the horns of a snail . . . The high grass warmed by the sun begins to give out a strong, heavy smell of honey. It will soon be midday, and Gerassim and Lubim are still floundering under the willow tree. The husky bass and the shrill, frozen tenor persistently disturb the stillness of the summer day.

'Pull him out by the gills, pull him out! Stay, I'll push him out! Where are you shoving your great ugly fist? Poke him with your finger – you pig's face! Get round by the side! Get to the left, to the left, there's a big hole on the right! You'll be a supper for the water-devil! Pull it by the lip!'

There is the sound of the flick of a whip . . . A herd of cattle, driven by Yefim, the shepherd, saunter lazily down the sloping bank to drink. The shepherd, a decrepit old man, with one eye and a crooked mouth, walks with his head bowed, looking at his feet. The first to reach the water are the sheep, then come the horses, and last of all the cows.

'Push him from below!' he hears Lubim's voice. 'Stick your finger in! Are you deaf, fellow, or what? Tfoo!'

'What are you after, lads?' shouts Yefim.

'An eel-pout! We can't get him out! He's hidden under the roots. Get round to the side! To the side!'

For a minute Yefim screws up his eye at the fishermen, then he takes off his bark shoes, throws his sack off his shoulders, and takes off his shirt. He has not the patience to take off his breeches, but, making the sign of the cross, he steps into the water, holding out his thin dark arms to balance himself . . . For fifty paces he walks along the slimy bottom, then he takes to swimming.

'Wait a minute, lads!' he shouts. 'Wait! Don't be in a hurry to pull him out, you'll lose him. You must do it properly!'

Yefim joins the carpenters and all three, shoving each other with their knees and their elbows, puffing and swearing at one another, bustle about the same spot. Lubim, the hunchback, gets a mouthful of water, and the air rings with his hard spasmodic coughing.

'Where's the shepherd?' comes a shout from the bank. 'Yefim! Shepherd! Where are you? The cattle are in the garden! Drive them out, drive them out of the garden! Where is he, the old brigand?'

First men's voices are heard, then a woman's. The master himself, Andrey Andreitch, wearing a dressing-gown made of a Persian shawl and carrying a newspaper in his hand, appears from behind the garden fence. He looks inquiringly towards the shouts which come from the river, and then trips rapidly towards the bathing shed.

'What's this? Who's shouting?' he asks sternly, seeing through the branches of the willow the three wet heads of the fishermen. 'What are you so busy about there?'

'Catching a fish,' mutters Yefim, without raising his head.

'I'll give it to you! The beasts are in the garden and he is fishing! . . . When will that bathing shed be done, you devils? You've been at work two days, and what is there to show for it?'

'It . . . will soon be done,' grunts Gerassim; summer is long, you'll have plenty of time to wash, your honour . . . Pfrrr! . . . We can't manage this eel-pout here anyhow . . . He's got under a root and sits there as if he were in a hole and won't budge one way or another . . .'

'An eel-pout?' says the master, and his eyes begin to glisten. 'Get him out quickly then.'

'You'll give us half a rouble for it presently if we oblige you . . . A huge eel-pout, as fat as a merchant's wife . . . It's worth half a rouble, your honour, for the trouble . . . Don't squeeze him, Lubim, don't squeeze him, you'll spoil him! Push him up from below! Pull the root upwards, my good man . . . what's your name? Upwards, not downwards, you brute! Don't swing your legs!'

Five minutes pass, ten . . . The master loses all patience.

'Vassily!' he shouts, turning towards the garden. 'Vaska! Call Vassily to me!'

The coachman Vassily runs up. He is chewing something and breathing hard.

'Go into the water,' the master orders him. 'Help them to pull out that eel-pout. They can't get him out.'

Vassily rapidly undresses and gets into the water.

'In a minute . . . I'll get him in a minute,' he mutters. 'Where's the eel-pout? We'll have him out in a trice! You'd better go, Yefim. An old man like you ought to be minding his own business instead of being here. Where's that eel-pout? I'll have him in a minute . . . Here he is! Let go.'

'What's the good of saying that? We know all about that! You get it out!'

'But there is no getting it out like this! One must get hold of it by the head.'

'And the head is under the root! We know that, you fool!'

'Now then, don't talk or you'll catch it! You dirty cur!'

'Before the master to use such language,' mutters Yefim. 'You won't get him out, lads! He's fixed himself much too cleverly!'

'Wait a minute, I'll come directly,' says the master, and he begins hurriedly undressing. 'Four fools, and can't get an eel-pout!'

When he is undressed, Andrey Andreitch gives himself time to cool and gets into the water. But even his interference leads to nothing.

'We must chop the root off,' Lubim decides at last. 'Gerassim, go and get an axe! Give me an axe!'

'Don't chop your fingers off,' says the master, when the blows of the axe on the root under water are heard. 'Yefim, get out of this! Stay, I'll get the eel-pout . . . You'll never do it.'

The root is hacked a little. They partly break it off, and Andrey Andreitch, to his immense satisfaction, feels his fingers under the gills of the fish.

'I'm pulling him out, lads! Don't crowd round . . . stand still . . . I am pulling him out!'

The head of a big eel-pout, and behind it its long black body, nearly a yard long, appears on the surface of the water. The fish flaps its tail heavily and tries to tear itself away.

'None of your nonsense, my boy! Fiddlesticks! I've got you! Aha!'

A honied smile overspreads all the faces. A minute passes in silent contemplation.

'A famous eel-pout,' mutters Yefim, scratching under his shoulder-blades. 'I'll be bound it weighs ten pounds.'

'Mm! . . . Yes,' the master assents. 'The liver is fairly swollen! It seems to stand out! A-ach!'

'The fish makes a sudden, unexpected upward movement with its tail and the fishermen hear a loud splash. . . they all put out their hands, but it is too late; they have seen the last of the eel-pout.

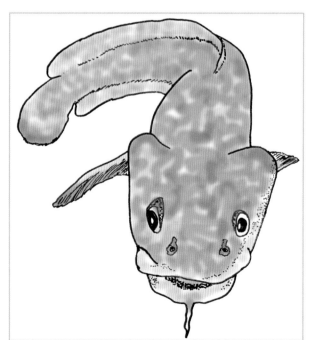

Figure 4.7
Russian writer and angler Anton Chekhov wrote a famous tale about the efforts of two Russian peasants trying to dislodge a burbot from submerged tree roots (© Dr Mark Everard)

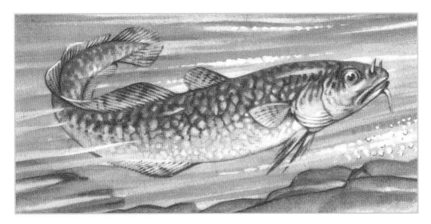

Figure 4.8 Burbot painting by E.V. Petts featuring as part of the 1960s set of 'Freshwater Fish' picture cards included by Brooke Bond in packets of their PG Tips and other tea and coffee products

Burbot have also been the subject of paintings. In the spring of 2001, the English artist Chris Gollon (born 1953) was invited to present a solo show by the *River and Rowing Museum* at Henley-on-Thames. Part of this show, inspired by the museum's collection of stuffed fish in cases and emulating Breughel's etchings, included a triptych titled *Big Fish Eat Little Fish* painted specially for the event. The centrepiece of the left panel of this set was a burbot, rising upwards from the muddy riverbed towards a frog at the water's surface.

A more widely known burbot painting was part of a series by the artist E.V. Petts illustrating a set of 'Freshwater Fish' picture cards, included by Brooke Bond in packets of their PG Tips and other tea and coffee products in the 1960s.

Perhaps the best-known of all burbot paintings is that of a fish '... *caught in the neighbourhood of Driffield*'. This specimen was immortalised by Alexander Francis Lydon (1836–1917), a British watercolour artist, illustrator and engraver of natural history and landscapes. A series of Lydon's wonderful paintings of British freshwater fishes illustrate Houghton's classic book *British Fresh-water Fishes* (Houghton, 1879). These paintings have since been widely used in many settings including on 'The Complete Angler' crockery range by the Welsh Portmeirion pottery company.

Figure 4.9 A.F. Lydon's painting of a burbot '. . . caught in the neighbourhood of Driffield', featuring in the Reverend W. Houghton's 1879 book *British Fresh-water Fishes*

In terms of music, the only songs which are known to refer to burbot have been made by Kalmah, a melodic 'death metal' band (if that is not an oxymoron) formed in 1998 in Oulu, Finland. Tracks by Kalmah which feature burbot include the 1998 demo record 'Under the Burbot's Nest', as well as the track 'Burbot's Revenge' from their 2003 album 'Swampsong'. Unlike the trout (inspiration of Franz Schubert's *Die Forelle*), the burbot seems not to have inspired any great works of classical music.

Burbot have doubtless graced the dinner tables of countless painters, authors, musicians and artists of other persuasions, but there is little evidence that the living fish has inspired a wider palette of great art!

Burbot and political history

Burbot have a place in political history, featuring highly in the protests relating to the drainage of Fenland. English antiquarian Sir William Dugdale (1662) published a monumental history of the drainage of the English Fens in *The history of imbanking and drayn-ing of divers fenns and marshes* . . ., describing in detail reclamation schemes around Norfolk in the preceding half-century. Within his extensive documentation, Dugdale (1662) reports that, despite the support of King Charles II, drainage works were halted for five years by,

> . . . reason of the opposition which diverse perverse-spirited people made thereto, by bringing of turbulent suits in law . . . and making of libellous songs to disparage the work.

One of these 'libellous songs' was *The Powtes Complaint*, now a traditional Norfolk song the authorship of which is lost in the mists of time. This protest song put voice to popular resistance to what was one of the most audacious land-grabs in English history. The vast mosaic of wetlands comprising Fenland was not only a wealth of biodiversity, including abundant burbot, but also the livelihood resource for a diversity of small communities making a living from the harvesting of fish and fowl as well as sedges, reeds, peat and other wetland vegetation.

For all of this natural wealth and resource, outsiders viewed these wild Fenland places as inhospitable, prone to winter floods and despised as a putrid, disease-infested quagmire. The native inhabit-ants of Fenland were considered equally wild and degenerate, primi-tive and debauched, and just as much despised and disparaged by higher and landowning classes.

In fact, the drainage of Fenland for 'agricultural improvement' was considered an aggressive dispossession from these people. It brought the formerly wild commons, supporting traditional liveli-hoods and cultures, into drained, fertile and privately owned farm-land reaping private profit. Passage of the Drainage Act in early 1601 created rights and incentives for investors – 'undertakers' as they were known – to make contracts with locals to receive a per-centage of profits from any land they 'reclaimed'. Soaring agricul-tural prices in the early 17th century led to massive drainage and enclosure of common land of many types, in this case the former moist habitats of Fenland and other burbot-rich and wider wetland habitats, to create more drained arable farmland. Privatisation of the commons obliterated the wetland ecology upon which many rural people depended for wild harvesting, grazing and summer farming, transforming the fens into lucrative monoculture. The degradation of the Fenland's ecosystems was as much a social as an environmental revolution, undertaken in the face of sabotage and other forms of violent resistance from people who had inhabited the living landscape for untold generations.

The *Powtes Complaint* was conceived as something of a battle-cry for this movement to preserve the last great bastion of wilderness in England. As such, it is one of the earliest and most important environmental protest poems in English literature, one of the likely anthems of enraged bands of local people protesting and sabotaging the dikes and pumps. It was not until works undertaken by Dutch engineer Sir Cornelius Vermuyden, commissioned by a consortium of wealthy investors spearheaded by Francis Russell, the Fourth Earl of Bedford, that more innovative methods resulted in completion of a substantial phase of drainage and with it the eradication of much of the region's biological and cultural diversity.

Whilst some historians assert that the powte, or pout, of the song's title was an old English word for sea lamprey, and others a generic name for a variety of fish in the cod family including cod and haddock, these are hardly species that would thrive in the diverse shallow freshwater marshes, pools and network of channels of the Fenland wetlands. The 'powte' was very probably the 'eel-powte', a 17th century name for what we call today the 'eelpout' or burbot. Transformation of the boggy wetlands into arable land was a death knell of its iconic species, the powte or burbot, sung in protest by those people whose lives and traditions shared the same needs and ultimate fate.

There are four surviving texts with slightly differing versions of The Powte's Complaint, the version here reproduced from Literary Norfolk (undated),

Come, Brethren of the water, and let us all assemble,
To treat upon this matter, which makes us quake and tremble;
For we shall rue it, if't be true, that Fens be undertaken
And where we feed in Fen and Reed, they'll feed both Beef and Bacon.
They'll sow both beans and oats, where never man yet thought it,
Where men did row in boats, ere undertakers brought it:
But, Ceres, thou, behold us now, let wild oats be their venture,
Oh let the frogs and miry bogs destroy where they do enter.
Behold the great design, which they do now determine,
Will make our bodies pine, a prey to crows and vermine:
For they do mean all Fens to drain, and waters overmaster,
All will be dry, and we must die, 'cause Essex calves want pasture.
Away with boats and rudder, farewell both boots and skatches,
No need of one nor th'other, men now make better matches;
Stilt-makers all and tanners, shall complain of this disaster,
For they will make each muddy lake for Essex calves a pasture.
The feather'd fowls have wings, to fly to other nations;
But we have no such things, to help our transportations;
We must give place (oh grievous case) to horned beasts and cattle,
Except that we can all agree to drive them out by battle.

Burbot and human health

The health benefits arising from eating fish are well known, as the flesh is low in fat and high in Omega-3 oils. As discussed previously, the oil from burbot also provides many of the health benefits attributed to cod liver oil as it has been known since the 1930s to have an extremely high vitamin A and D content.

However, there is also a potential 'dark side' to the inclusion of burbot meat and products in the diet. Burbot are known to carry spores of the 'broad tapeworm', *diphyllobothrium latum*. This tapeworm can thrive in the human intestine if the fish is eaten raw or only partly cooked, causing the disease diphyllobothriasis. *Diphyllobothrium latum* is native to Scandinavia, western Russia and the Baltic countries, though it is now also present in North America, especially the Pacific Northwest. The tapeworm does in fact infest a range of fish species, with no particular bias towards the burbot. Just make sure you cook your burbot (or other freshwater fish) well!

Burbot and local culture

Just as with great works of art, the burbot seems not to be at the epi-centre of a plethora of great contemporary cultural events. However, we have already noted the US 'Burbot Bash' held in Flaming Gorge Reservoir straddling the states of Wyoming and Utah, as well as the International Eelpout Festival on Leech Lake near the town of Walker in Minnesota.

Burbot then make their unique contribution to culture, but perhaps not to its higher echelons!

Trends in burbot populations

Burbot are not generally considered endangered across their wider geographical range. However, a scientific review paper published in 2010 titled *Worldwide status of burbot and conservation measures* (Stapanian et al., 2010) concluded that, although burbot are wide-spread and abundant throughout much of their natural range, many populations had already been extirpated or were endangered or in serious decline.

Across several European countries the burbot is considered sub-stantially threatened. The burbot became extinct in Britain around the early 1970s with the last confirmed specimen seen in 1969. In Belgium, burbot had been driven to extinction over the same approximate time period, though have more recently been the focus of a reintroduction programme.

Burbot numbers are also declining in a number of US rivers, in many instances for reasons that are unclear.

Burbot conservation issues across the world are explored in more detail later in the book (see Everard, Chapter 5, this volume, 2021)

The disappearing British burbot

The nature of the disappearance of burbot from British waters remains to a certain extent mysterious. The burbot, along with the common sturgeon, is one of only two fish species known to have become extinct across Britain over recent centuries.

As addressed in the opening chapters of this book, burbot have always had a restricted distribution in Britain. In *The Freshwater Fishes of the British Isles*, Regan (1911) noted of the burbot,

> In Britain it seems to be confined rivers flowing to the North Sea, from Durham to Norfolk.

Houghton (1879) in *British Fresh-water Fishes*, also summarised his then current knowledge of burbot distribution in Britain as,

> In our own country the Burbolt is rather a local fish . . . According to Yarrell, the Nottingham market was, in his time, occasionally supplied with examples for sale. It is found in the rivers of Yorkshire and Durham, Norfolk, Lincolnshire, and Cambridgeshire.

Formerly, burbot were restricted to the rivers of eastern England, from County Durham in the north to the Great Ouse system in the east including the Great Ouse, Little Ouse, Cam, Thet and Waveney, also including the Trent, the Tame, Dove, Derwent, Neme, Skerne, Esk, and Foss rivers. This restricted distribution was related to the former Doggerland land bridge with contemporary continental Europe, as previously discussed (see Everard, Chapters 1 and 2, this volume, 2021).

Incidentally, in his review of burbot distribution, Marlborough (1970) noted a report featured in the May 15th, 1964 edition of the journal *Fishing* by a Mr P Dumbill of Warrington, of a 38 cm burbot that had been taken from the River Tame in Greater Manchester by a Mr G. S. Norris in 1880, and then lodged in a local museum. This was reportedly the only known record of burbot from the North West of England (Malborough, 1970). However, it is entirely possible, and indeed more likely, that this location might have been mistaken for the River Tame in the West Midlands, a tributary of the Trent and part of the proven former range of the burbot.

In his book, *The Art of Angling*, Mansfield (1957) suggested that burbot had already started to become scarce by 1900. Later in the century Wheeler (1969) in *The Fishes of the British Isles and North West Europe* stated that,

> Its status in British waters is not clear. Records available suggest that it is very much more rare than it was formerly, but there is little objective evidence at hand to indicate its abundance in previous centuries.

In *Key to British Freshwater Fishes* Maitland (1972) noted that there were only six recorded captures between 1960 and 1972 in the United Kingdom (UK). The last confirmed capture of a burbot from a UK river was on the 14th of September 1969, taken from the Old West River in Aldreth, Cambridgeshire. Sporadic unsubstantiated reports of captures and rumours of relic populations continue. Some reported yet unsubstantiated sightings on the rivers Eden, Esk and Great Ouse have been mentioned already in this book. Another reported capture of a burbot was relayed to me during the drafting of this book. This time the reported location was Loch Treig in the Highlands of Scotland, a location well outside the natural range of the species though otherwise suitable in habitat terms. However, further investigation of this reported capture suggested it was a hoax. Despite these persistent and unsubstantiated rumours, it is generally accepted that the species became extinct from the British Isles most likely in the early 1970s.

As with many environment problems, it is rare for there to be just a single cause. The fact that lake burbot commonly spawn in mid-winter, commonly under ice, should give us some clue that they are adapted to cooler conditions. There is some speculation that British rivers simply became too warm for burbot, burbot then outcompeted by other better-adapted fishes. However, the Western European clade of burbot has become adapted to smaller rivers and has a higher thermal tolerance than the lake strains. The wider distribution of the Western European strain of burbot in continental Europe at similar and more southern latitudes rather argues against thermal stress as a primary cause of the loss of genetically similar British burbot. Another possibility is that the general nutrient enrichment of the eastern rivers by agricultural, industrial and residential intensification has played its part. These pressures have also caused considerable sedimentation of river and lake gravels which may further compromise the spawning success of burbot along with general ecological health.

If we are looking for the biggest 'smoking gun', the wholesale drainage of land and channelisation of rivers, disconnecting or eliminating floodplain wetland habitat and radically changing flow regimes, are far more likely primary causes. More spate-driven flow regimes in river channels, and the eradication of flooding in riparian meadows for the matter of months essential for the incubation of eggs and larvae, are strongly implicated in preventing the effective reproduction of burbot.

Burbot are also vulnerable to changes in nutrient concentrations, both from the enrichment of waters with nitrogen but also from the construction of dams causing nutrient starvation in river systems by the slowing of sediment and nutrient flows. Competition between the larval stages of the relatively few fish species found in nutrient-poor lakes is known to be an important factor affecting their survival.

The devastating impacts of the application of synthetic pesticides and other problematic agrochemicals will not have helped the situation, though were probably secondary causes driving declines that significantly preceded post-Second World War land use intensification. We can also rule out excessive predation, at least in British waters, as the progressive decline of the burbot preceded by some years the pesticide-driven elimination of otters from many eastern rivers from the 1950s and the resurgence in inland cormorant populations from the 1970s. Nevertheless, the collapse of burbot and a range of other fish species in the North American Great Lakes between the 1930s and 1960s has been attributed in part to a population of explosion of sea lamprey (*Lampetra marinus*), compounding the pressures of over-exploitation, habitat degradation and declining water quality including the accumulation of newly invented and widely applied pesticides.

The mystery of the vanishing British burbot remains. Although the primary causes seem clear, the wider causes are inevitably multifactorial. However, as we will cover in greater detail subsequently in this book (see Everard, Chapter 6, this volume, 2021), there are plans to explore the potential reintroduction of a species that has sometimes been dubbed 'the dodo of the rivers' into British waters.

Burbot as a perceived conservation problem

Some conservation concerns relate not to declining natural burbot populations, but to places where burbot may pose threats. These threats relate to ecosystems into which burbot have been introduced beyond their native range. As with any species, any organism moved beyond the native range and ecosystems with which it coevolved has the potential to become a problematic alien invasive. As the burbot is principally a predator, it can pose a threat when introduced accidentally or deliberately into new waters.

One such locality where the burbot is perceived as a problem alien invasive species is the Green River drainage in the US states of Utah and Wyoming, following the apparently illegal introduction of this fish in the mid-1990s into the Big Sandy reservoir. The burbot has since been perceived to have had an adverse effect on native fish populations and sport fisheries more widely in the upper Colorado River Basin. Most of the Green River drainage has sequentially become invaded by this fish upstream of the Flaming Gorge Dam in Utah, with indications of further spread downstream. Owing to concerns about the potential impacts of the predator on the high-quality brown trout fishery and the sockeye salmon population in Flaming Gorge Reservoir, the Utah Division of Fish and Game instituted 'no release', 'catch and kill' regulations for the burbot in Utah waterways from 2010. Not only have these regulations been found to be largely unenforceable, the Green River Chamber of Commerce in the US state of Wyoming has also established the annual 'Burbot Bash' on Flaming Gorge Reservoir, driven in part to cull the burbot population. Events such as these increase the popularity of angling for the species. That, along with the known cultural and economic benefits of burbot for the region, hardly create conditions conducive for policies seeking to eradicate the fish.

The consequences of the spread of non-native burbot are difficult to determine as so many other factors affect fish populations across the global range, not least the impacts of the large dams holding the reservoirs into which the fish have been introduced. But, as a guiding principle, any introduction of alien fish and other species poses potentially severe risks and must be prevented for future security.

A further significant risk is posed if burbot from a genetically differing strain are introduced into waters hosting a specific, locally adapted strain with associated lifestyle adaptations. In any restocking or reintroduction programme, it is essential that burbot of correct local genetic provenance are selected. There are, for example, populations in which burbot can grow up to 90 cm. The introduction of burbot from these populations into French or Belgian rivers, where native Western European strains usually do not exceed 50 cm, could have devastating impacts through their predatory and notably cannibalistic behaviour.

A concluding thought

Here's an interesting concluding thought. Given their northern distribution, association with deep and cool water, and habits that mean they will only very rarely be spotted from the bank, could the Loch Ness Monster in fact be a population of giant burbot?

No, I don't think so either! Since burbot have never been found in fish surveys of the lake, the deepest point of its largely uncharted bed is 230 m (755 ft), and the former recorded British distribution of the fish extended only as far north as Durham, it seems entirely implausible. Mind you, no huge and presumably air-breathing relic aquatic dinosaurs have shown up in fish surveys either and surely, of the two, the burbot is the more likely?

Perhaps the most important thing though is that, like the irrational drive that sees anglers occasionally embarking on expeditions to rediscover forgotten British populations of this cryptic fish, burbot retain an aura of mystery reminding us all that, as Shakespeare may have written were Hamlet a burbot angler,

There are more things in heaven and earth, Horatio, than are dreamt of in your philosophy.

Chapter 5

Burbot conservation

Burbot seem to be thriving across Siberia. In Finland, they are the most common freshwater fish found in glacial lakes, clean rivers and very substantial areas of low salinity habitat around the Baltic Sea coast. They are common too across interior Alaska and Canada. However, in areas of intense human activities and development, burbot seem to be faring far less well. In many places across their extensive natural range, burbot are declining and, in some, they have actually been driven to extinction.

In their research, Stapanian et al. (2010) found that, despite the variable fortunes of burbot across their global range, few regions were found to consider burbot in management plans. This was attributed to the lack of popularity of this fish for game and commercial purposes.

A canary in the coalmine

Owing to their exacting habitat, hydrological and water quality requirements, which is shared by many other species of plants, invertebrates, water birds, mammals and other fish and organisms, burbot are something of the metaphoric 'canary in a coalmine' in the indication of the health of integrated freshwater ecosystems. The viability of burbot populations is threatened by the physical, chemical or biological characteristics of any link in a chain of essential habitats, or the links between them. The same is also true of all of

Figure 5.1 Wild burbot in deep water in Khuvsgul Lake, Mongolia (© Batbold Tsedensodnom)

the other species dependent upon the same ecosystems that support the needs of this highly sensitive 'canary' fish.

The 'indicator' role of the burbot is one shared by other species such as elephants, tigers, the mahseer fishes of Asia, and other species that use linked habitat types to complete their life cycles. If populations of the indicator species are healthy, the many other species that also depend upon these habitats are also likely to prosper.

However, burbot differ from tigers, pandas, blue whales, sturgeon, Atlantic salmon and other more charismatic indicators. Indeed, they are often considered ugly, inspire little interest amongst anglers and promote no significant tourism. They are more often entirely disregarded as they frequently remain cryptic components of the ecosystems in which they occur. But this does not belittle their value as indicator species. If they decline or vanish, we know that our ecosystems and the diversity of their species, along with the many benefits that these ecosystems provide to us humans, are in trouble.

Burbot conservation

Burbot are not generally considered endangered across their wider geographical range. On a global scale, the International Union for Conservation of Nature (IUCN) Red List (the world's most comprehensive information source on the global extinction risk status of animal, fungus and plant species) assesses the burbot as being of 'Least Concern' (LC) (NatureServe, 2013), indicating that it is not perceived as a focus for species conservation. Also, under the Convention on International Trade in Endangered Species of Wild Fauna and Flora (CITES), an international agreement between governments aimed to ensure that international trade in specimens of wild animals and plants does not threaten their survival, the burbot is classed as 'Not Evaluated' and therefore not included in any of the three CITES Appendices of species requiring particular protection (CITES, UNEP-WCMC, 2020). These global assessments are founded on the fact that burbot remain common and widespread across many of the cooler regions of North America and Asia. However, they are nonetheless declining or already extinct in significant parts of this Holarctic range.

In the United States (US), no national nature conservation legislation (particularly the Endangered Species Act (ESA) and Multinational Species Conservation Act) explicitly covers burbot. The US Fish and Wildlife Service (2003) conducted a formal review of burbot in the lower Kootenai river in 2001 to determine whether that population should be protected under the ESA, though it ruled in March 2003 that federal protection was not warranted. However, a range of legislation relates to burbot across some of the US states in which the fish occurs. Some is unsupportive of burbot populations, such as a fishing pamphlet issued by the Department of Natural Resources in Wisconsin listing burbot as a 'rough fish' with no protections being provided such as closed seasons, bag limits or restrictions on methods of take. Washington, Oregon, Iowa and South Dakota also all lack fishing regulations for burbot. No special regulations are imposed on anglers fishing for burbot in Minnesota, though spring-loaded, self-hooking, tip-up rigs (used for a range of species caught by ice fishing) are illegal. The status of burbot in Michigan remains unknown because burbot are not sampled by regular fish surveys. In Alaska, some restrictive regulations on sportfishing have been put in place for adversely affected lake burbot populations, but none are enacted for river populations. In the state of Wyoming, concerns about burbot populations due to increasing angler exploitation and overharvesting of stocks led the Game and Fish Department to implement more restrictive regulations during the late 1940s through to the 1960s. These restrictions included shorter fishing seasons, lower creels limits, restrictions on minimum length and a closure of the special winter fisheries on lakes where burbot spawn. In Idaho, a burbot recovery programme is in place in the Kootenai River, the only place burbot occur in the state, following the decline of the species after the construction and operation of the Libby Dam in 1972.

Canadian law does not schedule burbot under wildlife legislation (principally the Endangered Species Act, Wildlife Act or Wilderness and Ecological Reserves Act). Neither are they specifically controlled under most fishery legislation (Fish Processing Licensing Board Act, Aquaculture Act, Fisheries Act, and the Professional Fish Harvesters Act). However, the exploitation of burbot through angling is controlled under the Maritime Provinces Fishery Regulations (Government of Canada, Undated), Part II of which relates specifically to burbot, including:

- 30.1 Gear Restrictions ('No person shall fish for burbot except by angling')
- 30.2 Close Times (imposing closed season for fishing in different catchments)
- 30.3 Quotas and Length Restrictions (which also vary by catchments).

In the western Canadian state of British Columbia, burbot is 'yellow-listed', meaning that the species is considered as not at risk. Sport fishing regulations in Alberta contain an exemption for 'wastage' of burbot, in which anglers are not legally required to retain harvested burbot for consumption.

In Europe, burbot are not listed under the European Commission

(EC) Habitats Directive, relating to the conservation of a wide range of rare, threatened or endemic animal and plant species as well as rare and characteristic habitat types. Neither are they listed under the Bern Convention, a binding international legal instrument covering natural heritage. Burbot are also listed as of 'Least Concern' (LC) under the 2011 *European Red List of Freshwater Fishes* (Freyhof and Brooks, 2011). Nevertheless, the European Union (EU) Species Action Plan for burbot was published in October 1999, reflecting the varying fortunes of the fish across its European range.

In the United Kingdom (UK), burbot enjoy the curious distinction of being both declared extinct, whilst also being given explicit protection across England and Wales (though Wales was never part of the historic range) under the Wildlife and Countryside Act 1981. This 1981 Act (as amended) is one of the UK's most significant pieces of nature conservation legislation, in which burbot are scheduled under Section 5 ('Animals which are protected') and Sections:

- 9.1 prohibition of killing or injuring
- 9.2 possession or control whether live or dead
- 9.4 protection from disturbance and intentional damage of their habitat
- 9.5 sale or advertising for sale whether live or dead.

Burbot are also listed in the UK as a Biodiversity Action Plan (BAP) species. The BAP programme was initiated following the 1992 Rio de Janiero 'Earth Summit'. However, until relatively recently, little action has followed the development of the burbot BAP.

There is a tendency for charismatic species to be prioritised for conservation attention. Fish tend to be overlooked, and particularly non-salmonids (fish other than trout and salmon). As burbot are often considered ugly, and are also hardly the most conspicuous of fishes given their cryptic behaviour, they are not only far away from being given the highest perceived priority, but their demise also tends to go largely unnoticed. An article published by the BBC (2015) titled *The decline of the 'disgusting' burbot* reflected this unfortunate set of largely unappealing features. The article quoted James

Maclaine, fish curator at the Natural History Museum in London, as saying that burbot are,

. . . fat looking and a little bit flabby and soft . . .

accounting for why people tend to have mixed feelings about the fish.

River rehabilitation for burbot recovery in Germany

A rehabilitation programme in the Lippe river in Germany is interesting in terms of how it has addressed the bottlenecks leading to a former precipitous decline in its burbot population. The Lippe is a river in North Rhine-Westphalia, Germany, a tributary on the right bank of the Rhine with a catchment area of 4.890 km². The Lippe was the focus of an extensive floodplain restoration programme commencing in 1997, with a reversal of the declining burbot population seen as an indicator of success.

Examination of historic literature had found that burbot were widely distributed throughout the Lippe catchment in the 19th century. However, excluding some sites into which artificial stocking had occurred, they had become extremely localised by the year 2000. Although adult burbot were making use of artificial habitats such as bankside riprap (rock or rubble armour put in place to protect banks from erosion), their breeding and nursery habitats were found to be extremely limited due to historic river straightening and the construction of embankments separating the river channel from its floodplain. Consequent bed scouring had deepened and further degraded habitat diversity within the river channel.

The characteristic spawning grounds used by burbot in cold conditions – shallows or tributaries including floodplain wetlands with still water or at best a gentle current and subject to low predation pressure – were found to be all but wholly eradicated. So too were any habitats that shared similar features with spawning habitats and would have been suitable for larval development by being rich in zooplanktonic food in spring, with low predation pressure and competition from other species. The pronounced lack of inundated floodplains, including backswamps and ponds as well as riverine shallows used by juveniles, was perceived as a bottleneck to burbot recruitment as well as compromising the ecology of the whole river system.

Restoration of the Lippe floodplain involved the collaboration of a range of project partners (the State of North-Rhine/Westphalia; Lippeverband; the City of Hamm LIFE projects; the District of Soest; Nordrhein-Westfalen-Stiftung Naturschutz, Heimat- und Kulturpflege; and Arbeitsgemeinschaft Biologischer Umweltschutz im Kreis Soest e.V). A range of connected habitat restoration works took place across the catchment. Formerly narrow and over-deepened sections of river were widened and raised, creating shallower and more diverse channel habitats, whilst also restoring historic sinuosity and increasing overall river length. Embankments were removed, providing the material to shallow the river channel and

Figure 5.2 A restored reach of the Lippe catchment, Germany, with a widened and shallowed channel reconnected with reinstated floodplain wetland habitats after removal of embankments (© Johan Auwerx/INBO)

space for channel widening. Critically, this enabled rehabilitated river channels to be reconnected with their floodplains, on which lakes and flood channels were also restored. Between 1997 and 2014, around 16 km of the river had been restored.

Prior to restoration, the burbot population overwhelmingly comprised adult fish. Responding to habitat diversification, young-of-the-year and juvenile burbot numbers increased rapidly. Stoll et al. (2017) conducted a study in which they gathered data from electrofishing over 21 consecutive years to research the fish-community responses in the Lippe River. The study involved gathering the data over a period of four years prior to restoration and a period of seventeen years following it, as well as the examination of multiple restored and unrestored reaches. The findings of the study showed that fish abundance peaked in the third year after restoration to a level six times higher than before the restoration. The species richness and abundance subsequently stabilised at two and three-and-a-half (respectively) times higher levels relative to the pre-restoration level.

Life-history and reproduction-related traits were found to best explain differences in how different species responded to restoration. Opportunistic and short-lived species perhaps unsurprisingly initially colonised the new habitats most rapidly. Assessment of traits of target species was found to be important for restoration outcomes. The particular breeding and nursery needs of burbot are a case in point, emphasising the vital importance of providing an appropriate matrix of habitats in any restoration or recovery programme.

Burbot restocking in Germany

Burbot populations in Germany are threatened by a range of pressures. Germany has a federal structure, comprising 16 Bundesländer (federal states) that have considerable autonomy, resulting in widely differing fishery policies existing between federal states. In Bavaria, approximately half of angling licence fees are directed into restocking threatened fish species. There are five priority species for restocking in Bavaria: huchen (*Hucho hucho*), grayling (*Thymallus thymallus*), barbel (*Barbus barbus*), nase (*Chondrostomus nasus*) and burbot. Burbot fingerlings have been captive bred and stocked in Bavaria since the last decade of the 20th century by the Bavaria Fisheries Association.

In 2004, and running through to 2010, EU funding supported a project to improve the keeping of burbot broodstock, captive breeding and larval growth through to the taking of dry feed by juvenile fish, and potentially then growth onwards to market size. Hendrik Wocher had a long history of working on burbot rearing in Bavaria, including work under this EU-funded project. From 2012, he set up the LOTAqua fish hatchery (see Box 5.1) as a private venture specialising in the production of burbot fingerlings. The vast bulk of the burbot fingerlings produced by LOTAqua are used for restocking, with perhaps one-tenth of them being sold to commercial fish farms for growing on as human food. Interest from sport fisheries in restocking with burbot is increasing year-on-year. LOTAqua takes special care to make sure that stock from appropriate genetic strains are made available for the restocking of different river systems.

Recovery of burbot populations in the United States

Burbot are also the subject of a recovery programme in the US state of Idaho, where wild populations have been declining,

A synthesis of the burbot stock history and future management goals in the Kootenai river in the state of Idaho by Hardy and Paragamian (2013) found that this fish is endemic to the river. Here, it once provided an important winter fishery used by indigenous people and European settlers. The population was substantial; the stock in the river and Kootenay Lake in British Columbia (north of the border in the Canadian state of British Columbia) comprising possibly the most robust burbot fisheries in North America.

However, the fishery in Idaho rapidly declined and collapsed in Kootenay Lake and the Kootenay river after the construction of the Libby Dam in 1972, culminating in the closure of the fishery in 1992. Operation of the Libby Dam for hydroelectric power generation

Box 5.1   The LOTAqua burbot hatchery, Germany

The LOTAqua hatchery operation, established and operative by Hendrik Wocher, is based in Überlingen on the northern shore of Lake Constance (the Bodensee) in the federal state of Baden-Württemberg near the border with Switzerland.

Maintaining genetic provenance is important for burbot conservation. Genetic differences are known between burbot from the Rhine and Danube catchments. For this reason, LOTAqua holds Bavarian broodstock used as a source for stocking the Danube catchment, kept isolated from Lake Constance broodstock the progeny of which are used to restock the Rhine catchment.

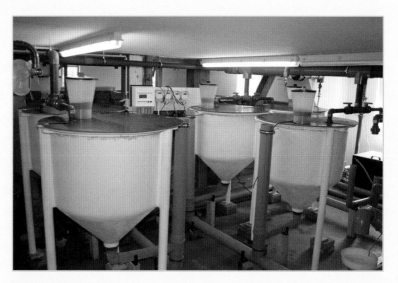

Figure 5.3 Burbot growing tanks at the LOTAqua hatchery, southern Germany (© Hendrik Wocher)

and flood control was found to have created major changes in the nutrient concentration, temperature and seasonal flow fluctuations of the Kootenai River, particularly in winter during the burbot's spawning period.

The Hardy and Paragamian (2013) study also outlined a conservation strategy to rehabilitate the Kootenai river's burbot population to a self-sustaining level. In particular, the strategy highlighted the need for operational discharge changes at the Libby Dam during winter to restore suitable temperatures and discharge conditions for the migration and spawning of burbot.

Owing to low and declining burbot numbers in the river system, coordination of culture and rearing of burbot by the Kootenai Tribe of Idaho, the British Columbia Ministry of Environment, the University of Idaho's Aquatic Research Institute, and the Idaho Department of Fish and Game (IDFG) was recognised as being important for the restoration of the species. Ultimately, sustainable levels of angling would need to be instigated, though only after the decline of the burbot population had been halted and reversed. The IDFG has contributed to research in the hatchery, and is also responsible for using Passive Integrated Transponder (PIT) tagging of fish for survival assessment to assess the outcomes of stocking at different locations according to the times at which different life stages were released.

Beard et al. (2017) reported on a study using PIT antennae on Deep Creek, a tributary of the Kootenai River, to evaluate movement of juvenile PIT-tagged burbot. 85% of PIT tags were relocated within 1 km of a release location, suggesting poor dispersal from stocking locations. Initial survival in the first seven months after release was low at 27%, though 63% of remaining burbot survived thereafter.

Burbot reintroduction

Another conservation strategy used where burbot, or other species, have been extirpated is the reintroduction of the species into parts of its former range. Species reintroduction is a complex, and sometimes contentious issue. It is, however, one with a successful recent track record. This includes in the UK where, over recent years, white-tailed sea eagles (*Haliaeetus albicilla*), Eurasian beavers (*Castor fiber*) and the large blue butterfly (*Phengaris arion*) are three examples of formerly extirpated animal species that have been successfully reintroduced to some of their former range.

The complexities associated with species reintroductions are the topic of the *IUCN Guidelines for Reintroductions and Other Conservation Translocations* (2013). These IUCN guidelines offer the following definition of 'reintroduction' as being,

> . . . the intentional movement and release of an organism inside its indigenous range from which it has disappeared.

The IUCN guidelines are detailed, documenting best global practice and many precautionary measures. Key aspects of the IUCN guidelines include understanding where the species originally occurred, as favoured locations for reintroduction. The IUCN guidelines also specify the identification of pressures explaining the decline and loss of the species, and how these pressures have been alleviated. It is also necessary to be clear about the purposes of reintroduction, and to ensure that no harm occurs to any possible remnant population in the reintroduction site nor to the donor site. The genetic provenance of potential donor stock and its compatibility with that of extirpated stock also has to be proven. Finally, the feasibility of achieving the reintroduction goals needs to be demonstrated.

The burbot has been successfully reintroduced into some localities across Belgium, southern Netherlands (the Beerze river) and Germany where it had formerly been driven to extinction.

Successful reintroduction of burbot to Belgium

A combination of pressures had driven burbot to extinction in Belgium, with the last confirmed capture of a specimen in 1957, though some scientists believe that final extirpation in Flanders (the Dutch-speaking northern portion of Belgium) was around 1970. This mirrors the timeframe of English extirpation, apparently due to similar cumulative causes.

Johan Auwerx of the Flemish institution *Instituut voor Natuur-en Bosonderzoek* (INBO) described to me the background to the burbot reintroduction programme in Belgium. From the 1960s to the

1980s, the rivers of Flanders had largely become fish-free, blackened by increasing water pollution. (The rivers in Wallonia, in the south of Belgium, were not quite so severely affected.) Belgium's rivers had also been substantially canalised for agricultural purposes in the 1960s, robbing them of connected wetland habitat on their floodplains. Pollution control programmes, initiated throughout Flanders in the early 1990s, began to clean up the rivers, resulting in their recolonisation by some fish species particularly including roach, gibel carp and sticklebacks. However, there was no repopulation by predatory fish species.

INBO initiated a series of species reintroduction programmes on behalf of the Flemish government's *'Agentschap voor Natuur en Bos': Agency for Nature and Forest* (ANB) from 1999 to redress this lack of key predatory fishes. Brown trout were identified as the most appropriate for upper rivers, pike for still waters and lower rivers, and burbot for lowland rivers. These species reintroduction programmes were, and still are, driven by conservation and not commercial concerns.

The Belgian burbot reintroduction programme was established with the objectives of achieving survival and subsequent breeding of individuals released into the wild (Worthington et al., 2008). This goal is shared with captive breeding programmes for other native species conducted at INBO's Center for Fish Farming, based in Linkebeek in the Belgian province of Flemish Brabant just south of the capital, Brussels. (INBO's native species breeding programme has progressively expanded to now include dace, chub, bullheads, crucian carp as well as two native toad species. A bitterling breeding programme has been discontinued, and there are plans to breed weather loach from 2022).

The Flemish burbot reintroduction project followed the IUCN guidelines with a feasibility study that included analysis of genetics and habitat suitability and the development of a captive breeding program to produce sufficient numbers of offspring to establish a wild population. This was then followed by the reintroduction of burbot larvae and juveniles into the wild, with post-release monitoring.

Prior genetic studies had determined that there were four

principal and distinct genetic strains of burbot across Europe: Western European, Northern European (Scandinavian), Baltic and Central European. Burbot formerly extirpated from Belgium were of the Western European strain, wild populations of which remain in some of the rivers of northern France. INBO research demonstrated some aspects of how genetic traits differ. When compared to the other strains, the Scandinavian strain of burbot was found to need water which was colder by about 1°C for breeding, but also was adapted as a lake strain that may not have fared so well had they been introduced into the smaller rivers inhabited by the Western European strain. These findings were important, as a great deal of prior European research had been conducted on the Scandinavian strain meaning that knowledge generated could not automatically be transferred to conditions in the west of Europe.

INBO initiated the captive breeding phase of the burbot reintroduction programme in 1999 under the instruction of ANB. The first step involved procuring burbot broodstock from wild populations in Austria in 1999 to learn how to keep and culture them. The INBO captive breeding program took place at INBO's specialised Linkebeek fish conservation hatchery, using an innovate recirculation aquaculture system enabling close monitoring and control of water parameters, optimising the environment for fish culture and growth. Learning about the specific needs of the burbot led to major improvements in hatchery techniques, with progressively increasing success rates in the survival of burbot larvae.

Once protocols for reproduction had been developed and tested, the conservation breeding programme progressed from 2005 using wild fish of Western European genetic provenance. These broodstock fish were sourced from the Meuse and Seine rivers of northern France, both of which are largely interconnected at their headwaters near the German border.

In addition to harvesting broodstock, wild burbot populations in these French rivers were assessed using PIT tags and larval surveys. Water quality and habitat features were also characterised. INBO's multi-year captive breeding and research programme underlined the fundamental importance of diverse and suitable habitat in these Western European rivers.

Figure 5.4 Inundated meadows on the River Meuse in northern France, providing natural breeding and nursery habitat for burbot (© Johan Auwerx/ INBO)

Burbot eggs spend long periods drifting and settling before hatching out larvae that are both tiny and substantially immobile, reliant on still or sluggish water as they float initially with the aid of their yolk sacs and embedded oil droplet and subsequently their swim bladders after they inflate. The relatively static margins of still waters inhabited by lake populations of burbot, typical of much of the Scandinavian range and in many other parts of their wider geographical extent, are ideal for the burbot's largely immobile egg and early larval stages. By contrast, Western European river populations of burbot depend on diverse static and sluggish wetland features across large, seasonally inundated floodplains, which also host rich planktonic food but few predators.

The information collected by INBO on the habitat requirements of Western European burbot during this programme was used to develop habitat suitability maps for juvenile and adult burbot, and to find areas suitable for their spawning and nursery needs. As a result of the research, three potential pilot reintroduction sites were identified in Belgium.

A further challenge to address was the tendency of multi-generational captive breeding to reduce genetic diversity. To reduce this tendency, INBO grows on burbot larvae and juveniles from different spawnings in ten different natural ponds. Broodstock from different ponds is mixed to maintain genetic diversity, which is analysed roughly every five years to ensure that diversity is maintained. The INBO broodstock was refreshed with new wild-caught French burbot in 2012. However, to avoid pressure on wild stocks in France, and also due to the daunting paperwork that has to be completed for wild capture and then for international transfer of stock for aquaculture, constant and frequent renewal of broodstock with wild fish is not a sustainable pathway for the captive breeding and reintroduction programme.

In the spring of 2005, more than two million cultured burbot larvae were reintroduced into several tributaries of the Grote Nete river in the Province of Antwerp, and the Bosbeek river in the Province of Limburg. This initial reintroduction was thought to have failed, as no juvenile burbot were recaptured during post-release monitoring. However, during the autumn of 2005, 2000 larger juvenile burbot (0+ age-class: juveniles towards the end of their first growing season) were released at several locations in the Grote Nete river, with a further 1000 released in the Bosbeek river. These stocks were regularly sampled by electrofishing. The sampling results showed that 42% and 12% of fish respectively of the second burbot stocking into the two rivers were recaptured in surveys, and these fish also showed good growth and condition. The Grote Nete river system, relieved of its former gross pollution and with suitable riparian habitat diversity, proved a particularly well-informed choice.

To ensure that adequate numbers of juvenile burbot were available for an effective reintroduction programme, INBO focused subsequently on a section of 25 km in length along the Grote Nete river with suitable water quality and habitat as a single reintroduction site. A habitat restoration project had also been conducted on the Grote Nete river, achieved through the collaboration of a consortium of institutions including ANB. The collaboration also included other public service bodies of the Flemish Government and non-governmental organisations such as Natuurpunt Beheer v.z.w (Natuurpunt, undated). This river restoration programme targeted 1850 hectares of the Grote Nete valley, aiming to restore a mosaic of habitats including sand dune and dune-heath vegetation, dry heath, species-rich grasslands, aquatic plant communities and forests on wet soils (Restoring Europe's Rivers, Undated). Restoration works involved clearing trees including the felling of 32 hectares of poplar plantations. River courses were also renaturalised, landscape hydrology improved and invasive water plants eradicated, to improve the spawning potential for fish in the river system. The works enabled succession in 42 hectares of former agricultural land and the restoration of 15 horticultural ponds with reintegration into the landscape. Remaining grazing regimes were also improved with nature conservation in mind. Downstream, and linked to this project, a renaturalisation programme was also undertaken on the De Kleine Hoofdgracht river to improve habitat for endangered fish species. These improvements included removing dikes and creating, or recreating, open water, marshland, oxbows and other floodplain habitat in a mosaic.

Juvenile burbot were stocked into the Grote Nete from INBO stock after one summer of captive growth in natural ponds, at which time they are 8–10 cm in length. However, some individuals attained a length of 20–25 cm over this time, which can create a problem for captive stock as they become distinctly cannibalistic at this size. The INBO rearing ponds were managed to provide the larval and juvenile burbot with a naturalistic planktonic diet, avoiding dried or other manufactured food, to ensure that the juveniles were conditioned to feed naturally once released into the wild.

During sampling in December 2007, sexually mature males and females in spawning condition were captured in the Grote Nete river demonstrating survival of the fish two years after reintroduction. Furthermore, new larvae were found during surveys in the winter in reaches where burbot had been reintroduced.

These positive results led to the subsequent release of juvenile burbot at other suitable locations in the Maarkebeek, Abeek and Ijse rivers. However, these additional introductions have since ceased, in order to ensure that sufficient numbers of burbot offspring are available to support the Grote Nete reintroduction programme.

Figure 5.5 The first larvae, at 4 mm, found in the wild resulting from natural reproduction of reintroduced burbot (© Johan Auwerx/INBO)

Determining the success of natural reproduction is difficult owing to the tendency of burbot to hide, making electrofishing largely ineffective and especially so for smaller individuals. Substantial variation in growth rate between individuals of this top predator during their first year of life complicates interpretation of their age. However, evidence of natural burbot reproduction is necessary to evaluate the success of a reintroduction. Innovative INBO research developed a visual inspection method to detect larval burbot in riparian zones. Direct visual inspection is possible on sunny, windless days in March, provided the necessary experience is available. However, in order to be independent of weather conditions, light traps were developed for use in stagnant, clear water bodies such as gravel lakes, depressions in floodplains, lakes and canals. Light traps were innovated and used from the beginning of March through to the end of April, considerably raising the chance of observations at locations

Figure 5.6 Johan Auwerx (INBO) placing a light trap to lure burbot larvae
(© Joachim Claeyé/AquaLota)

with low densities of burbot larvae. These methods were recommended for the assessment of natural reproduction.

Since 2014, larvae and juveniles originating from natural reproduction were found every year. Consequently, additional stocking in the upstream part of this river ceased in 2018. Burbot stocking campaigns are now taking place downstream, where major works have also been carried out to restore the river valley and create additional floodplain habitat.

Further genetic considerations

As discussed in the opening chapter of this book, there seem to be genetic differences even in closely located burbot populations. This was observed both in Germany as well as around Lake Michigan in the US, and is related to their philopatry (loyalty to particular locations), including for spawning or living out the rest of their adult lives.

Differences in genetic diversity were observed between major river basins, but also particularly between ostensibly adjacent lake and river habitats. As the INBO research had observed, lake-adapted Scandinavian strains have the benefit of relatively static still water margins enabling the survival of the tiny and relatively immobile eggs and larvae. By contrast, river populations need not only a diverse floodplain environment comprising mosaics of static and sluggish areas, but also the behavioural traits to exploit it including often long migrations by adult fish and an instinct to locate suitable headwater and floodplain habitat. This has implications for any stocking programmes and management methods used to preserve the genetic diversity and adaptive potential of burbot.

Burbot as a tool for biomanipulation

Biomanipulation is defined as the deliberate alteration of an ecosystem by adding or removing species, especially predators. Burbot have been found to serve this role where restocked.

The round goby (*Neogobius melanostomus*) is a member of the goby family (*Gobiidae*) naturally occurring in the Sea of Azov, Black Sea and Caspian basins. However, this fish has become a problematic alien invasive species where introduced beyond its native range. Adverse ecological impacts have been reported after introduction in a number of countries. This includes in continental Europe as well as in North America, where the round goby was reportedly accidentally introduced with ballast water from ships. The round goby also has adhesive eggs that are occasionally believed to be transported attached to the hulls of ships, facilitating introduction to other areas.

In Europe, the invasive round goby has become established in the Rhine catchment, as well as the Scheldt and Meuse in Belgium. In these catchments, the round goby has become the most abundant species, with predatory habits that appear to prevent the successful reproduction of native species such as zander (*Sander lucioperca*).

However, the habitat requirements of the round goby are similar to those of burbot, which are opportunistic predators that do not swim fast to pursue prey but rely on it coming close to them. Where restocking of burbot has occurred across Europe, it is reported that the biomass of round gobies has been substantially reduced. Stapanian et al. (2011) and Madenjian et al. (2011) reported similar findings from eastern Lake Erie in the US where invasive round gobies were first collected in offshore deep trawl assessments in 1999, the population size of these gobies increasing dramatically during 1999–2004. Round gobies were first found in the stomachs of burbot in 2001. By 2003 they had become the most important prey in the diet of burbot, implicated as a cause of larger burbot body sizes by ages 4–7 years and declining round goby numbers during 2004–2008. The adult burbot population was found to be feeding on round goby at an annual rate equal to 61% of the estimated round goby standing stock.

Figure 5.7 Burbot appear to play roles in the control of American signal crayfish, including where they have invaded European catchments (© Dr Mark Everard)

Burbot are also reportedly effective predators of another problematic and widespread alien invasive species, the American signal crayfish (*Pacifastacus leniusculus*). This crayfish has similar habitat needs to burbot, retreating into the cover of rough ground and tree roots. Therefore, this problematic crayfish is also readily encountered and consumed by burbot. Surveys from 1994 to 2006 in the Kootenai river in Idaho found that signal crayfish numbers, in some periods absent as by-catch, increased consistently as catch-per-unit-effort (CPUE) whilst burbot populations were diminishing sharply (Paragamian, 2010). We will return to the issue of the linked benefits of species reintroduction and its potential role in controlling problematic alien invasive crustaceans, in the following chapter (see Everard, Chapter 6, this volume, 2021).

Conservation co-beneficiaries?

Another species extirpated from British rivers since the 16th century is the Eurasian beaver (*Castor fiber*). This aquatic mammal was once widespread across Eurasia, but was hunted so close to near-extinction that, by the turn of the 20th century, only about 1200 beavers survived in eight relict populations. Progress has since been made with reintroductions across this range, including in Britain. Following both planned and accidental releases, a number of beaver populations have become established over recent years from Scotland to south west and southern England, and locally in various other regions of mainland Britain.

Eurasian beavers are 'ecosystem engineers' or, in other words, species that significantly modify, create, degrade or have other substantial impacts on habitats, and consequently on associated species richness and landscape-level heterogeneity. One of the principal activities of the dam-building and canal-creating activities of beavers is the proliferation of wetlands, diversification of flow regimes and sediment distribution in river valleys. These activities cumulatively improve habitat diversity and geomorphology, resulting in major benefits for biodiversity, water quality improvement, and hydrological buffering offering greater resilience against flooding and drought.

Diversification of functional wetland habitat in floodplains through the actions of beavers may be directly beneficial to burbot. This is particularly in respect of the critical winter-time spawning and larval development window, when still or sluggish inundated wetlands connected to river systems are essential for burbot to complete their life cycles. It is therefore likely, although unproven, that there are synergies between reintroductions of burbot and beavers. One or both reintroduced species may benefit, along with many other organisms and the wide range of ecosystem services provided by restored floodplain habitat in river valleys.

Chapter 6

Reintroduction of burbot to Britain

Although there has been a long-standing Biodiversity Action Plan (BAP) relating to burbot reintroduction from which little action had ensued, there has over recent years been renewed interest in reintroducing the burbot into parts of its former range in eastern England. Notwithstanding their long absence in the wild, burbot have not only appeared in Britain in recent years, but also spawned and produced young, albeit in captivity.

Any British burbot reintroduction programme would, for regulatory purposes, have to follow the requirements of the International Union for Conservation of Nature (IUCN) as published in the *IUCN Guidelines for Reintroductions and Other Conservation Translocations*. Demonstration of successful captive breeding and rearing of burbot is one of the building blocks of satisfying these IUCN guidelines that, as we have seen, also include identifying pressures explaining its decline and loss, and how these have been alleviated. It is also necessary to be clear about the purposes of reintroduction, and to ensure that no harm occurs to any possible remnant population in the reintroduction site nor to wild stock in the donor site. Finally, the feasibility of achieving the reintroduction goals needs to be demonstrated.

These factors are addressed in the following sections of this chapter, covering efforts undertaken by disparate players and sources interested in reintroducing the burbot to Britain.

Where could burbot be reintroduced in Britain?

Much of the overview of former British burbot distribution discussed in Chapter 2 (see Everard, Chapter 2, this volume, 2021) drew upon the research conducted by Tom Worthington during his PhD study. The topic of Tom's PhD was the feasibility of reintroduction of burbot to Britain, and his work was funded by the Trent Rivers Trust and the Esmée Fairburn Trust. This work has been highly informative about the former strongholds of burbot in England, and consequently potential candidate sites for their reintroduction to the wild. The funding sources also demonstrate a breadth of interests in the reintroduction of this fish.

As concluded in the overview of its former distribution in the United Kingdom (UK), the most likely places to find remnant burbot populations would be in the larger rivers in the eastern English counties of Yorkshire, Cambridgeshire, Norfolk and possibly East Anglia. It should be emphasised that no burbot have been found in successive surveys here since the 1970s. These locations then constitute the most likely places where burbot could feasibly be reintroduced.

Understanding factors behind the decline of British burbot

In the 2010 global review led by Martin Stapanian (Stapanian et al., 2010), discussed in the first part of this book, pollution and habitat change, and in particular the effects of dam building, were implicated as principal drivers of declining riverine burbot populations. In some river systems, burbot are known to migrate as far as 100–200 km for breeding, making any obstruction a serious impediment to them completing their life cycles. The impacts of invasive species were regarded as the principal reason for declines in some lake burbot populations, albeit compounded by warmer water temperatures. Both reasons were considered to be due to the combined effects of dam discharges and climate change. Fishing pressure was dismissed as a likely causative factor.

Reviews of burbot distribution in Britain by Worthington et al. (2010a and 2011) addressed five principal factors considered potentially contributing to their decline: disease, overfishing, pollution, habitat loss, and climate change. The first two of these factors were not considered significant. There was no evidence of disease wiping out burbot populations. Furthermore, although they were formerly put to a range of human uses, burbot were not heavily exploited in Britain. This is consistent with the findings of the 2010 global review of burbot population distribution and trends led by Stapanian et al. (2010), which concluded that overfishing was not a major problem for burbot stocks globally.

Pollution was seen as an issue, not only globally but also explicitly in Europe and in Britain. Although burbot are not exacting in water quality requirements, they do need reasonably high oxygen concentrations (5–7 mg of oxygen per litre of water). Extreme industrial pollution had driven sections of the River Trent to become almost oxygen-free in the later Industrial Revolution and through to the 1960s and 1970s. In the more rural catchments of the other eastern rivers in which burbot were present, agricultural intensification as well as increasing sewage inputs from the rising rural population were major causes of nutrient and organic enrichment, suppressing oxygen concentrations. Furthermore, as a species most comfortable at depth in standing waters, the deoxygenation of lower layers of water, where burbot retreat from higher surface temperatures, also poses a direct threat.

Happily, much of the gross pollution that blighted British rivers in the middle of the last century has been substantially addressed. This has resulted from a combination of better treatment of effluent from industrial, urban and local settlement sources, as well as the movement of much heavier industry overseas as the British economy made a substantial transition towards service sectors. Agricultural pollution, particularly diffuse organic and nutrient inputs from increasingly intensive farming practices, remains a problem in terms both of pollution and encroachment on riparian habitats. However, generally, there have been widespread improvements in the quality of water in Britain's rivers.

Habitat loss poses a major threat to burbot populations.

Agricultural intensification and land drainage, particularly driven by the post-Second World War policy priority of domestic food security but also with a far longer-term decline of wetland extent, had radical impacts on habitat diversity in river systems. Straightening of rivers robbed them of their sinuosity and impoundments blocked connectivity along the river. Perhaps most significantly of all, the partition of river channels from their adjacent floodplains through channel and bank dredging, as well as erection of embankments, destroyed a wealth of wetland habitats including adjacent backwaters and pools. These river modifications collectively created drier, less complex landscapes denuded of wetland diversity, static and sluggish water areas and wider lateral connectivity with river channels. The channels themselves, reengineered from diverse ecosystems into simplified drains, suffered more 'flashy' episodic flows, unbuffered by adjacent floodplains and with a tendency to become over-deepened.

Habitat loss is particularly significant for the egg and larval stages of burbot, which require standing water in floodplain wetlands over a period of months. Loss of these critical habitat features thereby creates a potentially terminal bottleneck to burbot recruitment. As we learned in the first part of this book, burbot spawn at low temperatures in winter. Lake populations often do so under ice at between 1°C and 4°C, with Western European river populations doing so at slightly higher, though still cold, temperatures. Spawning adults move at this time into the shallow waters of tributary streams and still water margins and adjacent wetlands to broadcast eggs communally. A very high number of tiny eggs is released to drift in the water column before settling slowly to come to rest in voids or cracks in underlying substrates. Thereafter, incubation is slow, taking anything from 30 to 60 days to hatch depending on water temperature. The tiny, immobile larvae that emerge remain inert whilst they absorb their yolk over a period of between two and four weeks during which time they are totally helpless. After the yolk sac and oil droplet are consumed, burbot larvae then swim up from the substrate, migrating into surface water to live out a pelagic lifestyle feeding on plankton for a matter of months. Only then do burbot larvae metamorphose into their adult form, dropping to the bed and banks of the water. In river systems, larval and fingerling burbot have been found almost exclusively in small tributaries and adjacent wetlands habitats. Where this marginal wetland mosaic, rich in connected still and sluggish water habitats, has been swept away and is no longer connected to river channels, burbot recruitment becomes impossible.

The availability of suitable habitat is a prerequisite for the successful reintroduction of a species and its persistence throughout its full life history. A 2012 study used data derived from field studies on burbot in France and Germany to assess the feasibility of reintroducing burbot into rivers of its former native range in eastern England (Worthington et al., 2012). This study found that the percentage of tree roots, aquatic vegetation and flow types were important predictors of adult burbot abundance in the studied rivers, and in particular off-channel habitat such as wetlands and backwaters essential for spawning and larval life stages. Absence of these features may potentially limit successful reestablishment. Rivers in the central and southern regions of the burbot's former English distribution were found to have the highest potential to provide suitable habitat. The study recommended that potential reintroduction sites should be preferentially selected on the basis of having appropriate spawning and nursery areas.

Temperature rise creates challenges for cold-water species such as burbot. Up to the 1980s, the River Trent, a former stronghold of burbot, had the ignominy of having more power stations along its length than almost any other river in the world. Heated effluent from this high density of power stations resulted in substantially elevated river temperatures, the river steaming in cooler weather. This heated water, in addition to the legacy of chemical pollution in the river, was hardly conducive to a thriving burbot population. Overlain on other anthropocentric alterations of river temperature are the impacts of climate change, affecting all British rivers.

Despite assertions from some quarters that rising temperature was a principal factor explaining the elimination of burbot from British rivers, particularly relating to spawning temperatures, many burbot populations across the wide geographical range of the species occur in areas that do not freeze in winter. These occur, for example, in some burbot habitats in northern Italy and northern France. Although former British populations were towards the southern end of the global Holarctic range of the burbot, these fish are widespread in similar latitudes in western Europe (notably in the Rhine, Meuse and Seine catchments) and populations exist further south in Germany, France and northern Italy. Many burbot populations in continental Europe – in Denmark, Belgium, Germany and France as well as northern Italy – occur in otherwise similar conditions to those found in eastern England. Furthermore, the 1950s–1970s, over which time burbot were lost from Britain, spanned a generally cooler climatic phase, bringing into question whether temperature is a primary driver of burbot extinction.

Factors potentially implicated in the demise of burbot in Britain as well as Germany and Belgium are multiple and interactive, though elimination of suitable spawning and nursery habitat appears disproportionately the most significant. It is no coincidence that Atlantic salmon (*Salmo salar*) and sea trout (*Salmo trutta*) were also extirpated from rivers in the east of England at the same time as the burbot. Populations of these salmonid fishes were able to repopulate at least some of the eastern rivers from which they had been lost as they have sea-going life stages. By contrast, burbot are unable to tolerate fully saline conditions and so, once lost from catchments, remained so.

Why reintroduce the burbot to Britain?

There is rarely a single reason for reintroducing a species. In fact, it is essential that reintroductions address multiple reasons.

One of these reasons is inherent. Burbot are a natural part of the British fauna. They are also one of the last vertebrate animals driven to extinction in England, and this still within living memory. As discussed in previous chapters when considering conservation, fish in general – particularly 'ugly' and cryptic fish such as the burbot – tend to be overlooked as conservation priorities. Restoring the burbot to at least part of its former range is, to many people, seen as a duty rather than a desire.

There are also statutory reasons for restoring burbot. Interests at UK government level in the feasibility of reintroducing burbot into the rivers of its former English range are driven by a response to the international Convention on Biological Diversity, which imposes a duty to monitor genetic diversity and to conserve biological resources. The two relevant regulators in England – Natural England as conservation regulator and the Environment Agency as environmental regulator – share responsibilities for authorising any reintroduction, and the Centre for Environment, Fisheries and Aquaculture Science (CEFAS: the government's marine and freshwater science advisers and data analysts with some regulatory duties) is tasked with authorising the import of stock.

Burbot are also indicative – the 'canary in the coalmine' as discussed previously (see Everard, Chapter 5, this volume, 2021) – of habitat types that have clearly been under grave threat. Reinstatement of the burbot to the rivers of which they were natural inhabitants also enables them to become an indicator for wider conservation of wetland and river habitats. They are in essence 'ambassadors' on behalf of the wider diversity of bird, plant, invertebrates, fish and other wildlife dependent upon these habitats.

Burbot could also provide sport for recreational anglers. Though never seemingly a popular angling target, there is interest in their reintroduction as a new sport fish.

Consideration of key sociological factors, including the views and attitudes of the local population, is critical to the success of species reintroduction projects. Engagement of stakeholders and the development of consensus and understanding is essential if future bad feeling associated with reintroduction activities is to be avoided. A consultation exercise was undertaken to seek and integrate disparate views from a wide range of stakeholders – amongst them anglers, conservationists and the general public – about a potential reintroduction of burbot to UK rivers (Worthington et al., 2010b). A snapshot of public attitudes was sought through both an online survey and an associated questionnaire, targeting these key stakeholder groups. More than 90% of respondents to both the survey and questionnaire supported the reintroduction of burbot. Anglers and participants with prior knowledge of the species cited burbot as an angling opportunity. The consensus of views of the overall surveyed constituencies suggested that reintroduction of the burbot was deemed feasible on ecological and biological grounds. There was no major public opposition.

In concluding this consideration of the need for multiple reasons to introduce burbot, I am reminded of a request made to me some time shortly after the turn of the millennium from a fishery manager responsible for a famous set of angling lakes in a very large gravel pit complex in southern England. In the larger of these lakes, there had been an invasion and population explosion of the alien invasive American signal crayfish (*Pacifastacus leniusculus*). The manager of these large and deep lakes was keen to find a way of reducing the population of these aggressive and prolific newcomers, which were making things problematic and unenjoyable for his paying anglers. At the time, though the historic wild presence of burbot in the Thames valley was uncertain, it struck me that this was not only the kind of larger lake environment that might serve as a burbot reintroduction site, but also that these carnivores might well make inroads into the crayfish. This seemed to me a 'double whammy' of conservation outcomes – providing a location for a species identified as a priority for reintroduction whilst simultaneously managing ecological problems associated with an alien invasive species – also serving the wishes of the fishery manager and even, in time, possibly offering a new species for anglers. However, the regulators at the time had no interest in this scheme. Consequently, without the capacity to gain consent, funding or even moderate support, the multi-outcome idea in my mind could progress no further.

Averting harm in the reintroduction site

IUCN guidelines concerning avoidance of risk to native populations at reintroduction sites might, on face value, seem irrelevant for a location in which the species had been declared extinct in Britain. After all, the once readily caught burbot had not featured in authenticated angling catches since 1969, and there had been no sign of burbot in repeated fish surveys since the 1970s by fishery management bodies

(Water Authorities, succeeded by the National Rivers Authority and subsequently by the Environment Agency), despite sporadic unsubstantiated claims of burbot sightings and landings.

However, it was necessary to provide regulatory bodies with evidence that no remnant, undetected burbot remained in candidate rivers into which the fish might be reintroduced. For this purpose, the Norfolk Rivers Trust commissioned an eDNA survey on the Rivers Wissey and the lower Great Ouse. (eDNA, or environmental DNA, comprises genetic material obtained directly from water samples, indicating the presence of the target organism somewhere upstream in the water body.) Perhaps unsurprisingly, results that became available from this survey towards the end of 2020 indicated no eDNA released by burbot in either river system (Norfolk Rivers Trust, as yet unpublished).

A further factor to consider under the criterion of causing no harm to native stock is to ensure that the genetic structure of donor populations is as close as possible, if not identical, to that which had existed in the reintroduction site.

Genetic analyses noted earlier in this book by Worthington et al. (2010a) highlighted the distinct lineage of preserved British burbot. However, its close relationship with the Western European clade highlights that extant stocks in northern France are most genetically appropriate for potential future reintroductions to English rivers. It is from this stock in northern France that the Flemish government's *Instituut voor Natuur-en Bosonderzoek* (INBO) harvested wild burbot for its conservation fish hatchery, which were used for the Belgian burbot reintroduction programme. Captive stock, one or two generations remote from wild harvested stock, have been identified as the most appropriate source for reintroduction to Britain. These fish have the benefit of having been health-checked with extensive biosecurity in place.

Biosecurity is essential to comply with regulations concerning the import of live fish, with extensive testing required to ensure that imported stock is free of certain specified pathogens. One such pathogen of particular concern is the monogenean fluke *Gyrodactylus salaris*, currently not present in the UK but a major concern for salmonid fishes across continental Europe. Although burbot have never been identified as a true host of this monogenean parasite, it is one of a range of parasitic organisms for which introduced stock must be proven to be uncontaminated.

Addressing both the need to avert risk at reintroduction and donor sites, as well as the potential for parasite transfer, burbot eggs may be the most portable and potentially sterile life stage to import to Britain for hatching and rearing for reintroduction.

Averting harm in the donor site

The Belgian sites to which burbot had been reintroduced, and the northern French stocks from which broodstock has been sourced, constitute the closest genetic match to specimens from British museums. Yet, low stock densities mean that wild harvesting from these locations is infeasible. Also, there is a risk of unintended transfer of parasites circulating in wild stock.

There is also a risk that tank-fertilised stock, even when reared in natural ponds, may not be as ecologically fit as individuals from wild populations, particularly if one or two generations distanced from wild stock. Nevertheless, obtaining stock from the secure and controlled INBO fish conservation hatchery is the best means to avert harm from the donor region. INBO takes precautions to ensure that the genetic diversity of its stock is maintained, and that high standards of biosecurity are ensured. Furthermore, as noted above, the transfer of eggs or fry from farmed populations offers not only the lowest impact but also the lowest risk of infection, particularly so if eggs are transferred as these can be effectively sterile.

Testing the feasibility of reintroduction

Amazingly, burbot have not only been brought into eastern England in recent years, but also bred there. This was initiated by the fisheries scientist Keith Easton, who convinced Ian Wellby, at the time working at Brooksby Melton College in Leicestershire, to attempt to breed some burbot. Ian, with the help of Keith, was able to persuade the college to invest in the import and secure keeping of burbot from around 2006. This was an important step to determine if it were possible not only to keep burbot in captivity but also to breed them in the ambient temperatures of the rivers into which they could conceivably be reintroduced. This burbot project became part of the teaching programme in the college, some aspects of the keeping and breeding of burbot forming part of student learning activities.

This breeding initiative was funded internally by the college and was achieved at low cost using donated tanks equipped with beer coolers to keep the temperature of the water down, supported by a growing body of interested observers. Consent to import and hold burbot was secured from CEFAS, and 19 wild-caught fish were subsequently imported into Brooksby Melton College from Denmark. To ensure biosecurity, including avoiding accidental import of fish parasites, a vet signed off the health of these fish before they left Denmark. The fish were quarantined and treated on arrival in secure holding facilities at Brooksby Melton College as a condition of the CEFAS consent.

These broodstock burbot adapted to captivity, though were strongly nocturnal in habit. They could not be persuaded to feed on anything other than live fish food. Ian Wellby was subsequently able to breed from a pair of these burbot, with artificial spawning, egg hatching and development of the emerging larvae to 'first feed' (when the larvae first ingest food having absorbed the yolk sac to which they are attached on hatching) at a temperature of 5°C, selected as representative of that of the River Trent in winter. Around 500 hatchlings were produced.

As a further biosecurity condition of the CEFAS consent, the captive burbot that had been held at the college for three years and their progeny were never meant to be released into the wild. When experiments had been concluded, the fry were destroyed humanely. The broodstock was sent to London Zoo, though none survived for long in this less controlled environment.

Regrettably, Brooksby Melton College has since stopped its fisheries programme. Ian Wellby has left the college to continue his activities as a private consultant, retaining a strong interest and involvement in burbot and their reintroduction.

Growing momentum for the reintroduction of burbot to Britain

As with many novel initiatives, continued interest in and development of ideas concerning burbot reintroduction has been driven by consortia of interested individuals. One such individual was Paul Kemp at Southampton University, who attracted funding from the Trent Rivers Trust and the Esmée Fairburn Trust for PhD research relating to whether the conditions were correct in British rivers for burbot to spawn successfully. The successful PhD candidate was Tom Worthington, whose work has informed substantial sections of Chapter 2 and other elements of this book.

A great deal of the initiatives cumulatively building towards a potential burbot reintroduction programme have been undertaken by disparate enthusiasts. The growing and strengthening consortium of interest in burbot reintroduction now includes greater attention from the statutory sector – Natural England and the Environment Agency – within which there had always been enthusiastic if isolated individuals promoting the work of others.

The Norfolk Rivers Trust reintroduction strategy

Since 2013, the Norfolk Rivers Trust has been working in partnership with a number of academic and conservation organisations to continue feasibility work and develop a potential reintroduction plan leading to a self-sustaining population of burbot in the UK. Taking on a focal role in bringing these initiatives together, the Trust produced a report *Reintroducing the burbot, Lota lota: a management plan* in 2020 (Norfolk Rivers Trust, as yet unpublished), funded by the Natural England Species Recovery Programme, bringing together a great deal of prior work (some of which is covered in this chapter). This reintroduction plan satisfied the criteria published in the IUCN guidelines (2013) which is a requirement for proceeding to regulatory consenting for a reintroduction programme.

The River Wissey is seen as offering the best prospects for burbot reintroduction. The Wissey is part of the catchment of the Great

Ouse, from which the last confirmed burbot capture occurred. The Wissey has the additional advantage of running within the geographical area over which the Norfolk Rivers Trust operates. Many of the pressures attributed to the extirpation of burbot from Britain have also either been reduced or are not as extreme in the Wissey as they are in many other former home range rivers. Limited fishing activities occur in the catchment, and water quality is generally good as the Wissey is naturally a chalk stream that is strongly spring-fed in the upper catchment. The Wissey has been assessed as having 'Good' chemical status and 'High' invertebrate status under EU Water Framework Directive criteria. Water quality in the Wissey is likely to have improved since the nadir of pollution in the 1950s–1970s due to substantial improvements to rural sewage treatment works. There is also a generally low level of diffuse agricultural pollution, with substantial investment in the past decade in addressing diffuse pollution across the catchment. Threats from sheep dip are also receding, as the formerly most damaging chemicals have been phased out. Overall, the threat of catastrophic pollution is perceived to have receded significantly, if not been entirely removed. As a spring-fed river, the temperature regime of the Wissey is also stable and generally cool, ensuring that the summer temperature can be kept to a reasonable level. Without this temperature moderation, there is a risk that the broodstock goes into too long a period of 'summer rest' and cannot produce high-quality eggs, which in turn compromises the survival of hatchlings. Populations of burbot occur in small, chalk-fed rivers in Denmark with strong similarities to the rivers of this part of Norfolk, and the INBO fish conservation hatchery at Linkebeek is also similarly spring-fed.

Habitat complexity and fragmentation are major pressures affecting the capacities of burbot to survive and breed, both in terms of longitudinal river connectivity as well as linkages with a mosaic of riparian wetland habitats across the floodplain. Habitats across the Wissey floodplain are generally reasonably well-connected, substantially aided by the fact that sizeable areas of the catchment are within a Ministry of Defence (MOD) training area and firing range. This affords potential reintroduction sites a high level of protection. The remainder of the catchment is also rural, with a low intensity of

Figure 6.1 Despite some historic channel straightening, connected wetlands on the floodplains of the River Wissey catchment provide habitat suitable for burbot reproduction (© Adrian Smith/Norfolk Rivers Trust)

farming. The Norfolk Rivers Trust has also been undertaking habitat restoration work in the catchment since around 2015.

Additionally, Environment Agency fisheries surveys demonstrate that there is a healthy fish population in the Wissey, including a mix of species as well as life stages. This suggests that spawning and nursery conditions in the catchment are in a largely healthy state, and that food sources potentially used by a variety of burbot life stages are present in abundance.

Sourcing of genetically compatible stock for reintroduction is an important consideration. The Norfolk Rivers Trust has good connections with the INBO fish conservation hatchery in Belgium from which, based on genetic analyses of British specimens held in museums, the closest genetic stock can be sourced. The Trust also has close links with the LOTAqua fish conservation hatchery in

Germany that may constitute an alternative source for reintroduction based on similar genetic compatibility.

There are issues still to be sorted at the time of writing, including finding a suitable secure British hatchery to hold and grow on stock, obtaining an import licence from CEFAS, and of course the ever-challenging need to secure funding.

There are also remaining uncertainties. For example, although the Wissey is otherwise rather similar to some of the Western European rivers in which burbot are present, have been reintroduced or where they have been restored, it is uncertain whether their success there is enabled by adult burbot exploiting other refuge habitats. For example, some of these continental European rivers have deeper, lower-salinity estuaries with the Baltic Sea where burbot may seek refuge, and they may just run rivers to spawn and grow on. The Wissey lacks deep water, and the estuary grades rapidly into full salinity which burbot cannot tolerate. However, the INBO programme of reintroducing burbot to the structurally similar Grote Nete river, combining with the Kleine Nete to form the Nete river that runs in turn into the Rupel river as a tributary of the Scheldt river with an estuary to the North Sea near Antwerp, has proven successful. This demonstrates that burbot reintroduction can succeed where there is adequate river habitat rather than a dependence on extensive, low-salinity estuarine and coastal zones. In fact, the re-established Grote Nete population is found to only migrate over limited distances, never reaching the brackish water area close to the North Sea which is also inaccessible to them due to barriers.

There is a hope that stocking burbot into the River Wissey can be initiated from late 2021, this phase running through to 2026 as burbot take up to two years to reach sexual maturity and to spawn successfully. It is hoped that this early phase can lead to natural recruitment at the selected reintroduction sites, with monitoring continuing for a further five years up to 2031.

Whether burbot reintroduction can succeed is, of course, uncertain. However, the attitude of the disparate network of interested champions is that we will not know until we try it.

On 31 March 2020, the Norfolk Rivers Trust hosted a webinar

communicating progress to that date. This webinar was recorded, and is freely available online (see the bibliography).

It's about a lot more than burbot!

One of the reasons why burbot make a fascinating and useful target for reintroduction is that they are cryptic, even considered ugly if not wholly disregarded, and are not a highly regarded angling quarry. In other words, the rationale for their reintroduction is not driven by their utility or public appeal.

More importantly, the diverse needs of burbot mean that they are indicators of the vitality of the catchment ecosystem for so many other species, as they are dependent not merely on good water quality but also on a range of interconnected wetland habitats. It is not just burbot but also this wider assemblage of species that benefits from reasonable or good water quality, temperatures that do not rise excessively and, above all, habitat diversity and connectivity both along the river and, most critically, across the floodplain. Burbot represent a flagship of a healthy river system supporting the needs of a huge diversity of other wildlife including invertebrates, plants, birds, mammals and different fish species.

More than that: we humans too benefit from healthy river systems in diverse ways, many of which have been formerly overlooked in our myopic eagerness to maximise food production and profitability. Food security is of course an entirely worthy aim. However, over a period of decades or centuries, we have thoughtlessly drained and obliterated complex riparian habitat with virtually no regard to its many beneficial functions. We have been guilty of being unquestioning about the wisdom of sweeping aside these multifunctional and multi-beneficial ecosystems. Protection or restoration of river corridors for the benefit of wildlife may, from a narrowly market focused view, look like an expense and a constraint for purely altruistic ends. Nothing could be further from the truth, as functional catchment habitat – the kind so extensively sacrificed on the altar of cheap commodity production – does so many wider things of immense cumulative value to society. These values include those stemming from natural hydrological functions contributing to the storage of water, buffering and building resilience against both flood and drought and potentially averting reliance on mechanical storage mechanisms and the damaging flooding of transport and other civil infrastructure. Intensive and diverse chemical processes in intact and functional floodplain habitat contribute to carbon storage, the cycling of nutrients and the improvement of water quality with many tangible benefits including averting costs of treating water abstracted for human uses. These habitats and processes naturally regulate human and livestock diseases, and host pollinators and the predators of pest organisms, whilst also promoting the regeneration of fish populations and other wildlife. Further values stem from medicinally and biochemically significant plants and wider wetland organisms. These wildlife and natural landscape features also potentially present ecotourism opportunities. Many less readily quantified values flow from the aesthetics of traditional and wild landscapes. This includes the sense of place they create and the spiritual meanings imputed to them. There is also the bequest to future human generations of this natural wealth and its many beneficial functions. Turning these thoughts over, these are the very values that we have serially overlooked and liquidated as inconsequential costs entailed in production of cheap farmed commodities and the sprawl onto naturally flat lands of urban and other infrastructure. These are the very values, we are beginning to realise, that constitute the foundational resources underpinning a truly sustainable future.

For all its cryptic and often unloved persona, the burbot is therefore a perfect emblem of a healthier river environment, beneficial to a huge diversity of native wildlife, not to mention human needs vital for our collective security and opportunity.

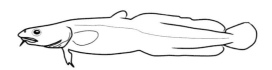

Chapter 7

Epilogue

Will I cast a fishing line for a burbot from a restored, self-sustaining British population in my lifetime? Will any of us? We simply do not know. But what we do know is that we will never know unless we try.

Reversing the tide of humanity's trajectory of ecosystem destruction, blighting the future for both wildlife and humanity, is an urgent priority. Repairing damaged life-giving rivers, including righting the myopic priorities of the past that ranked short-term

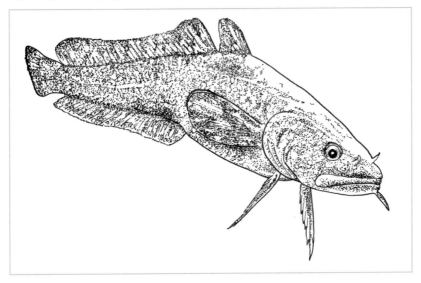

Figure 7.1 One day perhaps, burbot may swim and breed in English waters (© Dr Mark Everard)

food production over regenerative and healthy catchment ecosystems, is an absolute priority if we are to secure the wellbeing of all into the future.

The political and economic drivers of the past have rewarded the straightening of rivers and the drainage of floodplains and wetlands for narrow utilitarian purposes and generally short-lived benefits, ranging from food production to urban and industrial sprawl. But, what if we were to recognise that river systems and the landscapes of which they are part can confer wider and foundational forms of wealth? These include their capacities to store the carbon reserves that act as a bulkhead against a changing climate, and their role in buffering both flooding and drought, aiding adaptation to the impacts of climate change. What if we also valued the storage and purification of water within more intact catchment landscapes, reducing dependence on compensatory reservoirs as well as chemically and energetically intensive treatment of water extracted for public supply? What if we also valued the wide diversity of wildlife and landscapes that recycle nutrient chemicals, host pollinators and the predators of pest organisms, and provide cultural, aesthetic and spiritual meaning as well as a tourism resource and natural infrastructure providing a sense of place? Natural systems, in an intact and functional state, confer a breadth and depth of values to society without which a sustainable future – a future that is not robbed by short-term landscape desiccation that systematically erodes soil and degrades vital ecosystems – is all but impossible. It is not, of course, a case of 'either/or'. The need to grow food and find places to live, to work, and to produce the things we need is legitimate and necessary. The challenge is to innovate ways of living that work and develop in harmony with crucial natural habitats and their multiple functions and associated benefits.

And how do we know that our rivers are healthy, functioning in ways that support the needs both of diverse wildlife but also us humans, now and into the future? How do we know that our rivers can sustain our needs for fresh water and soil regeneration, nutrient and carbon cycling, buffering of floods and droughts, and a diversity of wildlife supporting our spiritual, recreational, wider cultural and other needs? We know because species that depend upon good water quality, diverse and connected habitats, and stable temperature regimes and hydrology are able to thrive in ecosystems at something closer to the natural state within which they co-evolved. Given critical needs across their life cycle, burbot are one such key indicator of our progress, or our lack of it. Their restoration or successful reintroduction are goals that can form the foundations of the rebuilding of a rich associated set of functions, values and meanings.

Though I am a Westcountry boy, I seriously hope that I see self-sustaining burbot populations restored to at least part of their natural range in eastern England. This is for their own inherent sake, as well as for the fact that it is incumbent upon us all to help bring back a species that our oversight drove to extinction within living memory. Importantly, it will also be a signal that, finally, humanity is turning the corner towards recognising the central importance of healthy ecosystems, underwriting wellbeing for us and all our activities coincident with the myriad species with which we share this world, both now and into the future.

My thanks to all those who have shared their knowledge to enable me to complete this book. Also, to those who have driven momentum towards reintroduction of the burbot to Britain and elsewhere. Many of these are the same people. Maybe, one day, we can cast a line together – be that a fishing line or a purely metaphorical one – into wilder waters by way of celebration come the day that burbot are once again there to be caught!

Bibliography

Alaska Department of Fish and Game, Sport Fish Division. (Undated). *Burbot Recipes*. Available at: https://www.adfg.alaska.gov/static/fishing/pdfs/sport/byarea/interior/publications/BurbotRecipes.pdf (accessed 31 March 2021).

Alaska Department of Fish and Game, Sport Fish Division. (Undated). *Filleting your burbot*. Available at: https://www.adfg.alaska.gov/static/education/angler/pdfs/filleting_your_burbot_tutorial.pdf (accessed 31 March 2021).

Angling Times. (2010). 'Extinct' burbot spotted in River Eden and Great Ouse. *Angling Times*, 9th July 2010.

Ashton, N.K., Cain, K.D., Hardy, R.S., Jensen, N.R., Ross, T.J. and Young, S.P. (2019) Temperature and Maternal Age Effects on Burbot Reproduction. *North American Journal of Fisheries Management* 39(6), 1192–1206.

BBC (2015) The decline of the 'disgusting' burbot. Available at: https://www.bbc.co.uk/news/magazine-33410217 (accessed 31 March 2021).

Beard, Z.S., Quist, M.C., Hardy, R.S. and Ross, T.R. (2017). Survival, Movement, and Distribution of Juvenile Burbot in a Tributary of the Kootenai River. *North American Journal of Fisheries Management* 37, 1274–1288.

Blumstein, D.M., Mays, D. and Scribner, K.T. (2018) Spatial genetic structure and recruitment dynamics of burbot (*Lota lota*) in Eastern Lake Michigan and Michigan tributaries. *Journal of Great Lakes Research* 44(1) 149–156.

Bosveld, J., Hendriks, J., Kranenbarg, J. and Lenders, H.J.R. (2015)

Historic decline and recent increase of Burbot (*Lota lota*) in the Netherlands. *Hydrobiologia* 757(1), 49–60.

Bruce, T.J., Cain, K.D., Jones, E.M., LaFrentz, B.R., Ma, J. and Oliver, L.P. (2020) Isolation and experimental challenge of cultured burbot (*Lota lota maculosa*) with *Flavobacterium columnare* and *Aeromonas* sp. isolates. *Journal of Fish Diseases* 43(8), 839–851.

Buckland, F. (1881) *The Natural History of British Fishes*, Society for Promoting Christian Knowledge, London, England.

Burbot Bash (2021) Ice tips for catching Burbot. Available at https://www.burbotbash.com/ (accessed 31 March 2021).

Carson, P. (1978) *The Fair Face of Flanders*, E. Story-Scientia, Belgium.

Chekhov, A. (1885) Налим (The burbot). *Peterburgskaya Gazeta*. July 1885.

CITES, UNEP-WCMC. (2020) *The Checklist of CITES Species Website. Appendices I, II and III valid from 28 August 2020*. CITES Secretariat, Geneva, Switzerland. Compiled by UNEP-WCMC, Cambridge, UK. https://www.cites.org/eng/app/appendices.php (accessed 31 March 2021).

Clemens, H.P. (1951) The Food of the Burbot *Lota Lota Maculosa* (LeSueur) in Lake Erie. *Transactions of the American Fisheries Society* 80(1), 56–66.

Clow, B. and Marlatt, A. (1929) The Antirachitic Factor in Burbot-Liver Oil. *Industrial and Engineering Chemistry Research* 21(3), 281–282.

Cocker, Mark. (2018) *Our Place: Can We Save Britain's Wildlife Before it is Too Late?* Penguin Random House UK, London, England.

Cott, P.A., Gunn, J.M., Hawkins, A.D., Higgs, D.M., Johnston, T.A., Martin, B., Reist, J.D. and Zeddies, D. (2014) Song of the burbot: Under-ice acoustic signalling by a freshwater gadoid fish. *Journal of Great Lakes Research* 40(2), 435–440.

Couch, J. (1864) *A History of the Fishes of the British Isles*. Groombridge & Sons, London, England.

Dahlstrom, P. and Muus, B. J. (1971) *Collins Guide to The Freshwater Fishes of Britain and Europe*. 1st edn. Collins, London, England.

Dugdale, W. (1662) *The history of imbanking and drayning of divers fenns and marshes, both in forein parts and in this kingdom, and of the improvements thereby extracted from records, manuscripts, and other authentick testimonies*. Alice Warren, London, England.

Everard, M. (2021) *Burbot. Conserving the Enigmatic Freshwater Codfish*. 1st edn. 5M Books Ltd, Essex, England.

Fantom, L. (2020) R&D brings burbot back to Belgium. Available at: https://www.hatcheryinternational.com/rd-brings-burbot-back-to-belgium/ (accessed 31 March 2021).

Freyhof, J. and Brooks, E. (2011) *European Red List of Freshwater Fishes [online]*. Luxembourg: Publications Office of the European Union. Available from: https://ec.europa.eu/environment/nature/conservation/species/redlist/downloads/European_freshwater_fishes.pdf (accessed 31 March 2021).

Furgała-Selezniow, G., Kucharczyk, D., Kujawa, R., Mamcarz, A., Skrzypczak, A., Targonska, K. and Żarski, D. (2014) Food selection of burbot (*Lota lota* L.) larvae reared in illuminated net cages in mesotrophic Lake Maróz (north-eastern Poland). *Aquaculture International* 22, 41–52.

Gardunio, E.I., Amadio, C.J., Keith, R.M., Myrick, C.A. and Ridenour, R.A. (2011) Invasion of illegally introduced burbot in the upper Colorado River Basin, USA. *Journal of Applied Ichthyology* 27(1), 36–42.

Giles, N. (1994) *Freshwater Fish of the British Isles: A Guide for Anglers and Naturalists*. Swan Hill Press, Devon, England.

Government of Canada. (Undated) *Maritime Provinces Fishery Regulations (SOR/93-55)*. Government of Canada. https://laws-lois.justice.gc.ca/eng/regulations/sor-93-55/index.html (accessed 08 April 2021).

Hardy, R. & Paragamian, V.L. (2013) A Synthesis of Kootenai River Burbot Stock History and Future Management Goals. *Transactions of the American Fisheries Society* 142(6), 1662–1670.

Harrison, J. (1988) Ice Fishing, the Moronic Sport. In: McGuane, T. (ed.) *Silent Seasons: Twenty-one Fishing Stories*. Clark City Press, Livingstone, Montana.

Harzevili, A.S., Auwerx, J., De Charleroy, D., Dooremont, I., Quataert, P. and Vught, I. (2004) First feeding of burbot, *Lota lota* (Gadidae, Teleostei) larvae under different temperature and light conditions. *Aquaculture Research* 35(1), 49–55.

Houghton, The Rev. W. (1879) *British Fresh-water Fishes*. Volume 2. W. Mackenzie, Edinburgh, Scotland.

Howell, M. and Fletcher, R. (2020). All about the burbot. Available at: https://thefishsite.com/articles/all-about-the-burbot (accessed 31 March 2021).

IUCN (2013) Guidelines for Reintroductions and Other Conservation Translocations: The Reintroduction and Invasive Species Specialist

Groups' Task Force on Moving Plants and Animals for Conservation Purposes, Version 1.0. Available at: https://www.iucn.org/content/ guidelines-reintroductions-and-other-conservation-translocations (accessed 31 March 2021).

Juniper, T. (2013) *What has nature ever done for us?* Profile Books, London, England.

Juta, U., Tosney, J. and Wellby, I. (2020) *Reintroducing the burbot, Lota lota: a management plan [Webinar]*. Norfolk Rivers Trust. Available at https://www.youtube.com/watch?v=Wwm9iKT03oM (accessed 31 March 2021).

Lewandoski, S.A., Deromedi, J.W., Gerrity, P.C., Guy, C.S., Johnson, K.M., Skates, D.L. and Zale, A.V. (2017) Empirical estimation of recreational exploitation of burbot, Lota lota, in the Wind River drainage of Wyoming using a multistate capture–recapture model. *Fisheries Management and Ecology* 24(4), 298–307.

Literary Norfolk (undated) Powte's Complaint. Available at: https://www.literarynorfolk.co.uk/Poems/powte's_complaint.htm (accessed 31 March 2021).

Madenjian, C.P., Einhouse, D.W., Pothoven, S.A. Stapanian, M.A., Whitford, H.L. and Witzel, L.D. (2011) Evidence for predatory control of the invasive round goby. *Biological Invasions* 13, 987–1002.

Maitland, P.S. (1972) *A Key to the Freshwater Fishes of the British Isles with Notes on Their Distribution and Ecology*. Freshwater Biological Association, Ambleside, England.

Mansfield, K. (ed.) (1957) *The Art of Angling, Volume 3*. Caxton, London, England.

Marlborough, D. (1970) The status of the burbot *Lota lota* (L.) (Gadidae) in Britain. *Journal of Fish Biology* 2(3), 217–222.

Marston, R.B. (1914) An angler who likes fishing for burbot. *Fishing Gazette*, 69, 370.

Millennium Ecosystem Assessment Panel. (2005) *Ecosystems & Human Well-being: Synthesis*. Island Press, Washington, DC.

Montagne, P., Gottschalk, D and Escoffier, G.A. (1938) *Larousse Gastronomique*. 1st edn. Librairie Larousse, Paris.

NatureServe. (2013). *Lota lota*. The IUCN Red List of Threatened Species 2013: e.T135675A18233691. https://dx.doi.org/10.2305/IUCN.UK. 2013-1.RLTS.T135675A18233691.en. (accessed 08 April 2021).

Natuurpunt (undated) Ontdek onze Natuurgebieden. Available at https://www.natuurpunt.be/ (accessed 31 March 2021).

Norfolk Rivers Trust (As yet unpublished) *Reintroducing the burbot, Lota lota: a management plan.* Funded Report.

Palińska-Żarska, K., Bilas, M., Krejszeff, S., Kucharczyk, D., Nowosad, J., Trejchel, K. and Żarski, D. (2014) Dynamics of yolk sac and oil droplet utilization and behavioural aspects of swim bladder inflation in burbot, Lota lota L., larvae during the first days of life, under laboratory conditions. *Aquaculture International* 22, 13–27.

Paragamian, V.L. and Bennett, D.H. (eds.) (2008) *Burbot: Ecology, Management, and Culture. Symposium 59.* American Fisheries Society, Maryland, Idaho.

Paragamian, V.L. and Wakkinen, V.D. (2008) Seasonal movement and the interaction of temperature and discharge on burbot in the Kootenai River, Idaho, USA, and British Columbia, Canada. In: Paragamian, D.L. and Bennett, D.H. (eds.) *American Fisheries Society Symposium 59* 55–77.

Paragamian, V.L. (2010) Increase in Abundance of Signal Crayfish May be Due to Decline in Predators. *Journal of Freshwater Ecology* 25(1), 155–157.

Plot, R. (ed.) (1686) *The Natural History of Staffordshire*, 1st edn. Printed at the Theater, Oxford, England.

Regan, C.T. (1911) *The Freshwater Fishes of the British Isles.* 1st edn. Methuen & Co. Ltd., London, England.

Restoring Europe's Rivers (Undated) Case study: restoration of the lowland river system Grote Nete. Available at: https://restorerivers.eu/wiki/index.php?title=Case_study%3ARestoration_of_the_lowland_river_system_Grote_Nete (accessed 31 March 2021).

Richardson, J. (1861) Fish in arctic regions. *British Medical Journal* 2(30), 109

Rotherham, I. D. (2010) *Yorkshire's Forgotten Fenlands.* Wharncliffe Books, Barnsley, England.

Rotherham, I. D. (2013) *The Lost Fens: England's Greatest Ecological Disaster.* The History Press, Cheltenham, England.

Schultz K. (2000) *Fishing Encyclopedia: Worldwide Angling Guide.* John Wiley & Sons, Inc, Hoboken, New Jersey.

Scott, W.B. and Crossman, E.J. (1973) Freshwater fishes of Canada. *Bulletin of the Fisheries Research Board of Canada*, 184, xi, 1–966.

Stapanian, M.A., Evenson, M.J., Jackson, J.R., Lappalainen, J., Madenjian, C.P., Neufeld, M.D. and Paragamian, V.L. (2010) Worldwide status of burbot and conservation measures. *Fish and Fisheries* 11(1), 34–56.

Stapanian, M.A., Edwards, W.H. and Witzel, L.D. (2011) Recent changes in burbot growth in Lake Erie. *Journal of Applied Ichthyology* 27(s1), 57–64.

Stoll, S., Bunzel-Drüke, M., Haase, P., Höckendorff, S., Scharf, M., Tonkin, J.D. and Zimball, O. (2017) Characterizing fish responses to a river restoration over 21 years based on species' traits. *Conservation Biology* 31(5), 1098–1108.

Tolstoy, L. (1878) *Anna Karenina*. The Russian Messenger.

US Fish and Wildlife Service (2003) News Release: Federal Protection Not Warranted for Lower Kootenai River Burbot. Available at: https://www.fws.gov/news/ShowNews.cfm?ID=350AF4B9-768A-4F78-9BB648E8E13239CF (accessed 31 March 2021).

Van Houdt, J.K., Hellemans, B. and Volckaert, F.A.M. (2003) Phylogenetic relationships among Palearctic and Nearctic burbot (Lota lota): Pleistocene extinctions and recolonization. *Molecular Phylogenetics and Evolution* 29, 599–612.

Van Houdt, J.K.J., De Cleyn, L., Perretti, A. and Volckaert, F.A.M. (2005) A mitogenic view on the evolutionary history of the Holarctic freshwater gadoid, burbot (*Lota lota*). *Molecular Ecology* 14(8), 2445–2457.

Vávra, R. (2020) Fish and chaps: some ethnoarchaeological thoughts on fish skin use in European prehistory. *Open Archaeology* 6, 329–347.

Vught, I., Auwerx, J., De Charleroy, D. and Harzevilli, A.S. (2008) Aspects of reproduction and larviculture of burbot under hatchery conditions. In: Paragamian, D.L. and Bennett, D.H. (eds.) *American Fisheries Society Symposium 59*.

Walton, I. (1653) *The Complete Angler. The Contemplative Man's Recreation*. 1st edn. Richard Marriot, London, England.

Wells. A. L. (1941) *The Observer's Book of Freshwater Fishes of the British Isles*. 1st edn. Frederick Warne and Co. Ltd, London, England.

Wetjen, M., Schmidt, T., Schrimpf, A. and Schulz, R. (2020) Genetic diversity and population structure of burbot Lota lota in Germany: Implications for conservation and management. *Fisheries Management and Ecology* 27(2), 170–184.

Wheeler, Alwyne. (1969) *The Fishes of the British Isles and North West Europe*. 1st edn. Macmillan and Co., London, England.

Wong, A. (2011) Heavy metals in burbot (*Lota lota* L.) caught in lakes of North-eastern Saskatchewan, Canada. *Journal of Applied Ichthyology* 27(s1), 65–68.

Worthington, T., Coeck, J., De Charlroy, D., Easton, K., Kemp, P.S.,

Osborne, P.E. and Vught, I. (2008) *The reintroduction of the burbot to the United Kingdom and Flanders.* In: Soorae, P.S. (ed.) *Global reintroduction Perspectives: Reintroduction Case Studies from Around the Globe.* IUCN/SCC Reintroduction Specialist Group, Abu Dhabi. pp. 26–29.

Worthington, T.A., Howes, C., Kemp, P.S. and Osborne, P.E. (2010a) Former distribution and decline of the burbot (*Lota lota*) in the UK. *Aquatic Conservation Marine and Freshwater Ecosystems* 20(4), 371–377.

Worthington, T.A., Kemp, P., Osborne, P.E., Tisdale, J. and Williams, I. (2010b) Public and stakeholder attitudes to the reintroduction of the burbot, *Lota lota. Fisheries Management and Ecology* 17(6), 465–472.

Worthington, T.A., Easton, K., Howes, C., Kemp, P.S. and Osborne, P.E. (2011) A review of the historical distribution and status of the burbot (*Lota lota*) in English rivers. *Journal of Applied Ichthyology* 27(1), 1–8.

Worthington, T.A., Bunzel-Druke, M., Coeck, J., Dillen, A., Easton, K., Gregory, J., Kemp, P.S., Naura, M. and Osborne, P.E. (2012) A spatial analytical approach for selecting reintroduction sites for burbot in English rivers. *Freshwater Biology* 57(3), 602–611.

Worthington, T.A., Van Houdt, J.K.J., Hull, J.M. and Kemp, P.S. (2017) Thermal refugia and the survival of species in changing environments: new evidence from a nationally extinct freshwater fish. *Ecology of Freshwater Fish* 26(3), 415–423.

Żarski, D., Kucharczyk, D., Mamcarz, A., Sasinowski, W. and Targonska, K. (2010) The influence of temperature on successful reproductions of burbot, *Lota lota* (L.) under hatchery conditions. *Polish Journal of Natural Science* 25(1), 93–105.

Index